Gene Targeting and Embryonic Stem Cells

Gene Targeting and Embryonic Stem Cells

Alison Thomson & Jim McWhir (Eds)

Division of Molecular Biology, Roslin Institute, Roslin, UK

BIOS Scientific Publishers
Taylor & Francis Group

LONDON AND NEW YORK

© **Garland Science/BIOS Scientific Publishers, 2004**

First published 2004

A CIP catalogue record for this book is available from the British Library.

ISBN 1 85996 360 9

Garland Science/BIOS Scientific Publishers
4 Park Square, Milton Park, Abingdon, Oxon, OX14 4RN, UK and
29 West 35th Street, New York, NY 10001 2299, USA
World Wide Web home page: www.garlandscience.com

Garland Science/BIOS Scientific Publishers is a member of the Taylor & Francis Group.

Distributed in the USA by
Fulfilment Center
Taylor & Francis
0650 Toebben Drive
Independence, KY 41051, USA
Toll Free Tel: +1 800 634 7064; E-mail: taylorandfrancis@thomsonlearning.com

Distributed in Canada by
Taylor & Francis
4 Rolark Drive
Marborough, Ontario MIR 4G2, Canada
Toll Free Tel.: +1 877 226 2237; E-mail: tal_fran@istar.ca

Distributed in the rest of the world by
Thomson Publishing Services
Sherlton House
North Way
Endover, Hampshire SP10 5BE, UK
Tel.: +44 (0)1264 332424; E-mail: salesorder.tandf@thomsonpublishingservices.co.uk

Library of Congress Cataloging-in-publication Data
Thomson, Alison.
Gene targeting and embryonic stem cells/Alison Thomson, Jim McWhir.
        p.; cm.
Includes bibliographical references and index.
ISBN 1-85996-360-9 (pbk.: alk. paper)
1. Gene targeting.   2. Embryonic stem cells.
[DNLM: 1. Gene Targeting—methods.   2. Cell Differentiation—genetics.   3. Genetic
Engineering—methods.   4. Stem Cells—cytology. QZ 52 T482g 004] 1. McWhir, Jim. II. Title.
QH442.3. T48 2004
571.6′1--dc22
        2004009191

Production Editor: Andrew Watts
Typeset by Charon Tec Pvt. Ltd. Chennai, India
Printed and bound by Cromwell Press, Trowbridge, UK

Contents

Preface

Gene Targeting and Embryonic Stem Cells is a practical guide designed for those who are considering moving into these important and growing areas of scientific activity. The book begins with an introductory overview of gene targeting and stem cells that ties together the various technologies covered. The chapters that follow have been contributed by a panel of internationally recognized researchers who are among the leaders in their respective fields and we are very grateful for the time and effort that they have expended on your behalf. This volume provides the necessary detail to design a variety of targeting constructs and to perform gene targeting with murine embryonic stem (ES) cells as well as ungulate fetal fibroblasts and also describes some of the limitations and technical difficulties likely to be encountered along the way. It includes RNA interference strategies through to a detailed description of how to both culture, transfect and differentiate established human ES cells and isolate new human ES lines. It also has a chapter on human embryonal carcinoma cells that covers surface antigen analysis via flow cytometry and fluorescence activated cell sorting. Finally, it includes a chapter on nuclear transfer. We hope that you find it both useful and enjoyable.

Alison Thomson
Jim McWhir
April 2004

Abbreviations

AAV	adeno-associated virus		**ICSI**	intracytoplasmic sperm injection
BAC	bacterial artificial chromosome		**IGF**	insulin-like growth factor
bDNF	brain-derived neurotrophic factor		**IVF**	*in vitro* fertilization
bFGF	basic fibroblast growth factor		**LIF**	leukemia inhibitory factor
BMP4	Bone morphogenic factor 4		**MAPK**	mitogen-activated protein kinase
CNS	central nervous system		**MEF**	mouse embryonic fibroblasts
CFTR	cystic fibrosis transmembrane conductance regulator		**MEK**	MAPK/ERK kinase
			MES	murine embryonic stem cells
dpc	days post coitum		**MHC**	major histocompatibility complex
EB	embryoid body			
EC	embryonic/embryonal carcinoma (cells)		**NGF**	nerve growth factor
			NHEJ	nonhomologous end joining
EF1α	Elongation factor 1α		**NT**	nuclear transfer
EG	Embryonic germ		**NT3**	neurotorophin3
EGF	epidermal growth factor		**PARP**	poly(ADP-ribose) polymerase
EGFP	enhanced green fluorescent protein		**PCR**	polymerase chain reaction
			PD	population doubling
ERK	Extracellular-signal-regulated kinase		**PDGF**	platelet-derived growth factor
			PECAM1	platelet endothelial cell-adhesion molecule 1
ES	Embryonic stem			
FACS	fluorescence-activated cell sorting		**PGC**	primordial germ cell
			PKR	dsRNA responsive protein kinase
FGF	fibroblast growth factor			
FISH	Fluorescence *in situ* hybridization		**PrP**	prion-related protein
			RA	retinoic acid
GGTA1	α1,3-galactosyltransferase		**RISC**	RNA-induced silencing complex
GSK3	glycogen synthase kinase 3		**RT**	room temperature
HCMV	human cytomegalovirus		**SCF**	stem cell factor
HES	human embryonic stem cells		**ShRNA**	short hairpin RNA
HGF	hepatocyte growth factor		**SiRNA**	small interfering RNA
HIV	human immunodeficiency virus		**TC**	tissue culture
			TE	trophoectoderm (cells)
HPRT	hypoxanthine phosphoribosyl transferase		**TFO**	triple-helix-forming oligonucleotides
HSV	Herpes simplex virus			
ICM	inner cell mass		**UTR**	untranslated region

An overview of gene targeting and stem cells

Jim McWhir, Alison J. Thomson and David Russell

At first glance this might appear an odd grouping of techniques. Yet, at least in the special case of embryonic stem cells, gene targeting and stem cell culture are commonly undertaken in the same lab. To some of us, though we really know better, it seems as though murine embryonic stem (mES) cells were expressly designed for genetic manipulation. They do everything we ask of them: grow indefinitely, accept incoming DNA, maintain a relatively stable karyotype and then go on to make chimeric animals and to pass modified genes through the germ line. With the help of heterosis, and a tetraploid host embryo to provide the extraembryonic membranes, mouse ES cells can even regenerate an entire animal (1). These extraordinary properties have enabled the generation of precise genetic modifications in mice by gene targeting followed by germ line chimerism, allowing us to ask important questions about gene function and to model human genetic disease.

There is another side to ES cells that is not necessarily related to their genetic manipulation. They have the inherent potential to give rise to virtually all cell types, thereby providing tools for the possible treatment of metabolic or degenerative disease. Since the isolation of the first human ES (hES) cells in 1998 (2) there has been a dramatic increase in the number of new workers who are beginning to culture these cells for the first time. Sadly, we cannot yet say of human ES cells, even tongue in cheek that they were 'designed for genetic manipulation' – at the present time they remain difficult to work with. It is also unclear to what extent the various cell lines available differ, to what extent they adapt to changes in their culture regime and if such adaptations have functional or tumorigenic consequences. We hope that the protocols herein will provide a useful basis for the further refinement to culture systems that will be so necessary to resolve these issues and to achieve regulatory acceptance.

But why have we included a chapter on nuclear transfer? The historical context of this book is the post Dolly biology of challenged conventions. If the nucleus of an adult differentiated cell can support development to term and beyond (3), what else may be possible? So nuclear transfer is perhaps only one manifestation of this broader phenomenon of nuclear reprograming that may eventually offer access to all of the varied cell fates encoded within a single nucleus, effectively making all cells potential 'stem cells'. Adding gene targeting to this equation renders all genetic lesions repairable, in principle, in all tissues.

1.1 The murine embryonic stem cell system

Chapters 2 and 3 describe the use of murine ES cells in gene targeting and nuclear transfer respectively. ES cell lines were first isolated by Evans and Kaufman (4), and independently, by Gail Martin (5) in 1981. Their experiments were inspired by the properties of the stem cells of germ cell tumors known as embryonal carcinoma or EC cells (see Chapter 6). EC cells had been shown to contribute to somatic tissues when injected into the blastocoel cavity of murine preimplantation embryos, but rarely if ever, contributed to the germ line. It was evident that if that deficiency could be overcome by deriving similar cells directly from the embryo, it might provide a route to germ line modification.

The process of ES isolation in mice can be characterized as the serial disaggregation of the undifferentiated component of an embryonic explant (most commonly a 3.5-day blastocyst). In permissive strains this process eventually gives rise to ES lines at a frequency of up to 30% depending upon level of expertise. The unsuccessful explants succumb to differentiation, generally to trophectoderm, followed by senescence. The process of ES isolation may be enhanced in mice by ovariectomy followed by supplementation with progesterone to induce delay of implantation (4). Certain mouse strains are more difficult or impossible to obtain lines from using conventional techniques, but do yield lines under conditions in which the differentiating component of the explant is removed, either by microdissection (6), by selective ablation (7,8), or is forestalled by inhibition of the ERK/MEK pathway (9).

Control of the undifferentiated growth of murine ES cells is now relatively well understood, in large part due to the work of Hans Scholer and colleagues (10) and of Austin Smith and his colleagues (11,12), in unravelling the control of mES self-renewal and differentiation. Three transcription factors are involved in controlling mES fate *in vitro*. The same factors *in vivo* control the fate of the epiblast from which ES cells arise. Oct4 is required *in vivo* for regulation of cell fate during the first differentiation event leading to trophectoderm and *in vitro* for the self-renewal of ES cells (11). Stat3 activation by leukemia inhibitory factor (LIF) is also required to sustain mES/epiblast self-renewal (12,13), and Nanog (14) is subsequently required *in vivo* to maintain the epiblast, its down-regulation triggering the second differentiation event that leads to parietal/visceral endoderm.

1.2 Human ES cells

There are many reasons why the isolation of additional human ES lines remains a high priority. Foremost among these is the near certainty that existing lines could not obtain regulatory approval for clinical application due to the nature of their derivation. The use of undefined medium including serum and the requirement for coculture of embryonic explants on a feeder layer of murine embryonic fibroblasts means that existing lines would be treated for regulatory purposes as xenografts. Additional reasons include the desirability of banking hES lines of diverse major histocompatibility complex (MHC) haplotype in order to minimize the problem of graft rejection and the utility for research purposes of the derivation of lines carrying specific genetic lesions. Chapter 7 describes the methods currently and most commonly in use for the successful isolation of hES lines.

The excitement attached to hES cells arises from their potential application in medicine. How much of this potential can be substantiated? Undirected differentiation of hES cells arises simply by their removal from feeders or conditioned medium. The real challenge is to direct that differentiation and to purify large numbers of functional cell types or their progenitors, ultimately by specific induction of a single differentiation pathway in defined media. The protocols provided in Chapter 8 provide a basis for the further development of directed differentiation of hES cells.

Zhang *et al.* (15) reported a transition of embryoid bodies to neural rosettes, which are subsequently harvested by selective dissociation as free-floating aggregates that generate both neurons and glia. Reubinoff *et al.* (16) observed expression of neural markers when hES cells were maintained in culture without passage or replenishment of feeders and in the absence of basic fibroblast growth factor (bFGF). These cells were then replated in serum-free medium to form free-floating spherical structures similar to neurospheres isolated from adult brain tissue. Both groups performed engraftment studies into the lateral ventricle of neonate rats and showed that transplanted cells migrated into many brain regions. However, it is unclear if these cells are functional. In a third study (17), a wide variety of ES-derived neural cell types were demonstrated *in vitro* and shown to have action potential. A limitation of these techniques, is the absence of effective protocols for the derivation of enriched populations of the most clinically relevant cell types in the central nervous system (CNS) – dopaminergic neurons (Parkinson's disease), cholinergic neurons (Alzheimer's) and oligodendrocytes for repair of myelination defects.

Several groups have reported the enhanced generation of functional cardiomyocytes from hES cells after treatment with 5-aza-2'-deoxycytidine (18,19) (a demethylation reagent), enrichment by Percoll gradient separation (18), and treatment with retinoic acid (RA) (20). Human ES cells also respond to osteogenic factors, giving rise to mineralizing osteoblasts with a similar time course to that observed with marrow stromal cells (21). Spontaneous differentiation of hES cells includes cells with characteristics of insulin-producing β-cells (22). Although this is an encouraging first step, these putative ES-derived islet cells remain small numbers of cells interspersed among a mixed population. Hepatocyte differentiation from hES cells is important both in regenerative medicine and for drug and toxicity testing, and an expandable population of ES-derived cells with the features of hepatocytes has recently been observed following treatment with sodium butyrate (23). Endothelial cells have been purified from differentiating hES cells using antibodies to platelet endothelial cell-adhesion molecule 1 (PECAM1) (24) and were seen to form vessel-like structures in matrigel and hematopoietic precursor cells arise when hES cells are cocultured with the murine bone marrow cell line S17 or the yolk sac endothelial cell line C1620 (25).

The general features of human ES cells in the undifferentiated state are thoroughly described in Chapters 6–8 and are not fully reprised here. Suffice it to say that there are important dissimilarities with murine ES cells. For example hES cells are not responsive to exogenous LIF and do not express the SSEA1 epitope characteristic of murine ES cells. The control of hES self-renewal and differentiation is not as well understood as it is for mES cells. However, recent results may have provided a way to short circuit the very elaborate signal transduction pathways controlling ES fate in both species. Pharmacological inhibition of glycogen synthase kinase 3 (GSK3) was shown to be sufficient to sustain ES self-renewal for 7 days in the absence of additional cytokines (26). It remains to be seen if that effect is maintained with prolonged culture. If so, and if hES cells can be shown to be unaffected in other respects following this treatment (e.g. tumorigenesis) then it may soon be possible to grow hES cells for therapeutic applications in completely defined media.

Protocols for the introduction to hES cells of foreign DNA are described in Chapter 9. Genetic modification of hES cells will allow us to engineer cells with features that provide protection against tumorigenesis post engraftment, that promote angiogenesis or that provide essential metabolic function – a combination of gene and stem cell therapy. The expression of tissue-specific markers can assist the isolation, purification and amplification of specific cell types for engraftment. Although there are concerns about the use of genetically modified cells *in vivo*, this approach in ES cells may provide the opportunity to use modified ES cells to study the basic biology of progenitor cells. This in turn may allow their purification directly from wild-type populations including adult stem cells from a variety of sources.

1.3 Gene targeting and ES cells in other species

Gene targeting in livestock species can be applied to the efficient production of therapeutic proteins in the mammary gland. Here the concept is to introduce a transgene into the 5'UTR of a mammary-specific gene such as B-casein, in order to ensure high-level tissue-specific expression. Although there are also, in principle, applications for the improvement of agricultural traits and animal production there is not, in the present economic and social climate, a great deal of enthusiasm for their uptake. More immediate are the possibilities for modeling human disease. Although mouse models of disease have been informative, they have in some instances, such as the cystic fibrosis transmembrane conductance regulator (CFTR) knockouts, not presented with the main pathology of the human disease.

Chapter 4 describes gene targeting in sheep fetal fibroblasts. Although these cells have been used successfully to target several genes it remains the case that no gene has yet been targeted that is not also expressed in fetal fibroblasts and that the senescence of targeted fibroblast clones is problematic both for identification of targeted clones and for their subsequent use in nuclear transfer (27). If proven ES cells were available from these species and had similar properties to mES cells, then gene targeting would certainly be greatly advanced for applications in farm animals.

Chapter 7 describes a method of isolation of hES cells that may inspire adaptation for livestock. A critical feature of hES isolation, and possibly for ungulate species as well, is the sensitivity of early passage ES cells to method of disaggregation, a sensitivity not usually observed for mES cells. It will be interesting to determine if lessons learned with hES cells can be turned to useful advantage in the search for ovine, bovine and porcine ES cells. Alternatively, there is some evidence that failure to isolate ES cells in the rat may arise as a consequence in that species of down-regulation of Oct4 (28). If this is also true in farm animal species then strategies to trigger Oct signaling may be required.

1.4 Gene targeting

Chapter 2 describes in detail the numerous elegant approaches to genetic modification of mice that are provided by mES cells, while Chapter 4 describes the less robust application of gene targeting to fetal fibroblast nuclear donor cells. In this area there remains considerable scope for technical improvement. Failure to target unexpressed genes in fetal fibroblasts may be due to reduced availability of DNA as a consequence of chromatin structure at unexpressed loci. Alternatively, homologous recombination may occur at these loci, but the expression of selectable markers may subsequently be inhibited, again as a consequence of chromatin structure. One possibility therefore is to include elements such as insulators in vector design. Another is to include additional features designed to enhance the underlying frequency of homologous recombination, such as cotransfection with recombinases (29) or to transiently alter chromatin structure (e.g. 30).

Although we were unable to include a chapter in this volume, certain viral vectors provide advantages in gene targeting. Parvoviruses are non-enveloped viruses with an encapsidated, single-stranded, linear DNA genome of approximately 5 kb. The dependent parvovirus adeno-associated virus (AAV) requires a separate helper virus such as adenovirus for productive infection, and has been developed as a gene transfer vector (31,32). AAV vectors typically contain foreign DNA between the *cis*-acting viral terminal repeats, and the viral *rep* and *cap* genes have been removed.

AAV vectors can transduce cells by multiple pathways, including integration into nonhomologous host chromosomal sites (33), or persistence as linear and/or circular episomes (34). In addition to these gene addition pathways, AAV vectors can pair with homologous chromosomal sequences and introduce site-specific sequence changes in a gene-targeting process (35). In primary human fibroblasts, targeting frequencies of 1% can be achieved without selection, and frequencies over 70% are possible by selecting for a vector-encoded transgene (35,36). Of 50 targeted loci that have been sequenced (36–39; and unpublished observations) and except for one example of a complex rearrangement involving two target loci (37), the sequence changes were accurately introduced without secondary mutations.

A variety of site-specific chromosomal mutations have been produced with AAV vectors, including small (<20 bp) insertions, deletions, and substitutions, as well as larger (1.0–1.5 kb) transgene cassette insertions (36–38). For optimal results the mutation or transgene being introduced should be centrally located in the vector genome, and the length of homology with the chromosomal target site should be maximized (38). The high efficiency, high fidelity, and success with normal human cells make AAV-mediated gene targeting ideally suited for gene targeting in human embryonic stem cells. We have recently shown that these vectors also work in hES cells (Russell, Thomson and McWhir, unpublished data).

We now know that gene targeting can also be accomplished in human ES cells using conventional plasmid-based vectors (40). Unfortunately, as with our own AAV-based hES work, this result has come too late to allow for the inclusion of an hES targeting chapter in this volume. Nevertheless, Chapter 9 provides methods for the efficient introduction of DNA to human ES cells that are also applicable to targeting constructs.

Gene targeting can also be achieved indirectly, by targeting the transcript. The relatively new field of RNA interference has provided a powerful tool for assessing the consequences of the down-regulation of specific genes. The main hurdle to the application of RNAi in mammalian cells was the coincident activation in those cells of the interferon response, triggering a general shutdown of the cell's translational machinery. However, early embryonic cells including ES cells don't have an active interferon response and it seems that in other cell types this response can be bypassed by the use of small (<30 bp) molecules. Protocols for the application of RNAi in ES cells are provided in Chapter 5.

1.5 Nuclear transfer

Nuclear transfer (Chapter 3) is not the only example of experimental nuclear reprograming but it is certainly the most spectacular, resulting as it does in viable (though not always healthy) offspring. Other forms of experimental reprograming include the generation of hybrid cells (41), incubation of permeabilized cells in xenopus ooplasm (42) and the forced expression of so-called 'master genes' (43). The extent to which these alternative systems can be pushed remains unknown but offers a number of exciting challenges. The possibility that cell fate can be experimentally controlled has clear and enormous therapeutic potential.

1.6 Conclusion

In the history of science new technologies have sometimes led nonspecialists to imagine applications that are of little interest to the scientific community and that far exceed the possible. In HG Wells' 1896 novel 'The Island of Dr. Moreau', surgical techniques could be used to give human characteristics to animals, with the inevitable disastrous consequences. Were he around today Wells might have engaged similarly with the possibilities suggested by new techniques in stem cells, regenerative medicine and tissue engineering. It is inevitable that our ability to reprogram cells to adopt a different fate, or to replace them with other cells that are programed with that fate, will have limitations. Perhaps the excitement that attaches to this field of science arises precisely because those limitations are not presently apparent.

If 'reprogramming' simply implies an experimental change to the program, then it is a very general term. At the simplest level we can reprogram by altering primary gene structure (gene addition and gene targeting). At a secondary level we can reprogram by targeting the destruction of a transcript (RNAi). And we can reprogram by altering the epigenetic structure of chromatin (nuclear transfer). Working out the rules that lead to predictable, reliable and useful outcomes should keep us busy for some time to come.

References

1. Schwenk F, Zevnik B, Bruning J, *et al.* (2003) Hybrid embryonic stem cell-derived tetraploid mice show apparently normal morphological physiological and neurological characteristics. *Mol Cell Biol* **23** (11): 3982–3989.
2. Thomson JA, Itskovitz-Eldor J, Shapiro SS, Waknitz MA, Swiergiel J, Marshall VS, Jones JM (1998) Embryonic stem cell lines derived from human blastocysts. *Science* **282**: 1145–1147.
3. Wilmut I, Schnieke AE, McWhir J, Kind AJ, Campbell KHS (1997) Viable offspring derived from fetal and adult mammalian cells. *Nature* **385**: 810–813.
4. Evans MJ, Kaufman MH (1981) Establishment in culture of pluripotential cells from mouse embryos. *Nature* **292**: 154–156.
5. Martin GR (1981) Isolation of a pluripotent cell line from early mouse embryos cultured in medium conditioned by teratocarcinoma stem cells. *PNAS* **78**: 7634–7638.
6. Brook FA, Gardner RL (1997) The origin and efficient derivation of embryonic stem cells in the mouse. *PNAS* **94**: 5709–5712.
7. McWhir J, Schneike AE, Ansell A, Wallace H, Colman A, Scott AR, Kind AJ (1996) Selective ablation of differentiated cells permits isolation of embryonic stem cell lines from murine embryos with a non-permissive genetic background. *Nature Gen* **14**: 223–226.
8. Gallagher EJ, Lodge P, Ansell R, McWhir J (2003) Isolation of murine embryonic stem and embryonic germ cells by selective ablation. *Trans Res* **12**: 451–460.
9. Buehr M, Smith A (2003) Genesis of embryonic stem cells. *Philosophical Trans Royal Soc London Series B Biol Sci* **358** (1436): 1397–1402.
10. Scholer H, Hatzopoulos AK, Balling R, Suzuki N, Gruss P (1989) A family of octamer-specific proteins present during mouse embryogenesis: evidence for germline-specific expression of an Oct factor. *EMBO J* **8**(9): 2543–2550.
11. Nichols J, Zevnik B, Anastassiadis K, Niwa H, Klewe-Nebenius D, Chambers I, Scholer H, Smith A (1998) Formation of pluripotent stem cells in the mammalian embryo depends on the POU transcription factor Oct4. *Cell* **95**: 379–391.
12. Williams RL, Hilton DJ, Pease S, *et al.* (1988) Myeloid leukaemia inhibitory factor maintains the developmental potential of embryonic stem cells. *Nature* **336**: 684–687.
13. Niwa H, Burdon T, Chambers I, Smith A (1998) Self-renewal of pluripotent embryonic stem cells is mediated via activation of STAT3. *Genes Develop* **12**: 2048–2060.
14. Mitsui K, Tokuzawa Y, Itoh H, *et al.* (2003) The homeoprotein nanog is required for maintenance of pluripotency in mouse epiblast and ES cells. *Cell* **113**: 631–642.
15. Zhang S, Wernig M, Duncan ID, Brustle O, Thomson JA (2001) In vitro differentiation of transplantable neural precursors from human embryonic stem cells. *Nature Biotechnol* **19**: 1129–1133.
16. Reubinoff BE, Itsykson P, Turetsky T, Pera MF, Reinhartz E, Itzik A, Benhur T (2001) Neural progenitors from human embryonic stem cells. *Nature Biotechnol* **19**: 1134–1140.
17. Carpenter M, Inokuma MS, Mujtaba T, Chui CP, Rao MS (2001) Enrichment of neurons and neural precursors from human embryonic stem cells. *Exper Neur* **172** (2): 383–397.
18. Xu C, Police S, Rao N, Carpenter MK (2002) Characterization and enrichment of cardiomyocytes derived from human embryonic stem cells. *Circ Res* **91** (6): 501–508.
19. Kehat I, Kenyagin-Karsenti D, Snir M, *et al.* (2002) Human embryonic stem cells can differentiate into myocytes with structural and functional properties of cardiomyocytes. *J Clin Invest* **108**: 407–414.

20. Schuldiner M, Yanuka O, Itskovitz-Eldor J, Melton DA, Benvenisty N (2000) Effects of eight growth factors on the differentiation of cells derived from human embryonic stem cells. *Proc Natl Acad Sci USA* **97** (21): 11307–11312.

21. Sottile V, Thomson A, McWhir J (2003) In vitro osteogenic differentiation of human ES cells. *Clon Stem Cells* **5** (2): 149–156.

22. Assady S, Maor G, Amit M, Itskovitz-Eldor J, Skorecki KL, Tzukerman M (2001) Insulin production by human embryonic stem cells. *Diabetes* **50**: 1691–1697.

23. Rambhatla L, Chui C-P, Kundu P, Peng Y, Carpenter MK (2003) Generation of hepatocyte-like cells from human embryonic stem cells. *Cell Transplant* (in press).

24. Levenberg S, Golub JS, Amit M, Itskovitz-Eldor J, Langer R (2002) Endothelial cells derived from human embryonic stem cells. *Proc Natl Acad Sci USA* **99** (7): 4391–4396.

25. Kaufman DS, Hanson ET, Lewis RL, Auerbach R, Thomson JA (2001) Hematopoietic colony-forming cells derived from human embryonic stem cells. *Proc Natl Acad Sci USA* **19**: 10716–10721.

26. Sato N, Meijer L, Skaltsounis L, Greengard P, Brivanlou AH (2004) Maintenance of pluripotency in human and mouse embryonic stem cells through activation of Wnt signalling by a pharmacological GSK-3-specific inhibitor. *Nature Med* **10** (1): 55–64.

27. Thomson A, Marques M, McWhir J (2003) Gene targeting in livestock. *Reproduction* **S61**: 495–508.

28. Buehr M, Nichols J, Stenhouse F, *et al.* (2003) Rapid loss of Oct-4 and pluripotency in cultured rodent blastocysts and derivative cell lines. *Biol Reprod* **68** (1): 222–229.

29. Dominguez-Bendala J, Priddle H, Clarke A, McWhir J (2003) Elevated expression of exogenous Rad51 leads to identical increase in gene targeting frequency in murine embryonic stem (ES) cells with both functional and dysfunctional p53 genes. *Exper Cell Res* **286**: 298–307.

30. Dominguez-Bendala J, McWhir J (2004) Enhanced gene targeting frequency in ES cells with low genomic methylation levels. *Trans Res* **13**: 69–74.

31. Carter PJ, Samulski RJ (2000) Adeno-associated viral vectors as gene delivery vehicles. *Int J Mol Med* **6**: 17–27.

32. Rutledge EA, Halbert CL, Russell DW (1998) Infectious clones and vectors derived from adeno-associated virus (AAV) serotypes other than AAV type 2. *J Virol* **72**: 309–319.

33. Miller DG, Rutledge EA, Russell DW (2002) Chromosomal effects of adeno-associated virus vector integration. *Nat Genet* **30**: 147–148.

34. Duan D, Sharma P, Yang J, *et al.* (1998) Circular intermediates of recombinant adeno-associated virus have defined structural characteristics responsible for long-term episomal persistence in muscle tissue. *J Virol* **72**: 8568–8577.

35. Russell DW, Hirata RK (1998) Human gene targeting by viral vectors. *Nat Genet* **18**: 325–330.

36. Hirata R, Chamberlain J, Dong R, Russell DW (2002) Targeted transgene insertion into human chromosomes by adeno-associated virus vectors. *Nat Biotechnol* **20**: 735–738.

37. Inoue N, Hirata RK, Russell DW (1999) High-fidelity correction of mutations at multiple chromosomal positions by adeno-associated virus vectors. *J Virol* **73**: 7376–7380.

38. Hirata RK, Russell DW (2000) Design and packaging of adeno-associated virus gene targeting vectors. *J Virol* **74**: 4612–4620.

39. Zwaka TP, Thomson JA (2003) Homologous recombination in human embryonic stem cells. *Nature Biotechnol* **21** (3): 319–321.

40. Tada M, Tada T, Lefebvre L, Barton SC, Surani MA (1997) Embryonic germ cells induce epigenetic reprogramming of somatic nucleus in hybrid cells. *EMBO J* **16** (21): 6510–6520.

41. Collas P, Hakelien AM (2003) Teaching cells new tricks. *Trends Biotechnol* **21** (8): 354–361.

42. Odelberg SJ, Kollhoff A, Keating MT (2000) Dedifferentiation of mammalian myotubes induced by msx1. *Cell* **103** (7): 1099–1109.

Advances in gene targeting in murine embryonic stem cells

2

Alan R. Clarke

2.1 Introduction

The ability to introduce precise changes into the germ line of mice has revolutionized our approach to investigating gene function *in vivo*. Despite its wide-ranging use, the technology is still relatively young, with the first engineered mice only created in the late 1980s. Since that time the experimental approaches used have rapidly evolved, and this chapter addresses some of the more recent advances in this technology. The simplest use of gene targeting has been the inactivation of a given allele in order to allow loss of function studies, and much effort has been focused on enhancing the efficiency of this process and in generating high-throughput protocols for gene inactivation. A second area of activity has been in exploring the potential of gene targeting to investigate gain-of-function and conditional mutations, and to precisely engineer chromosomes. Taken together with the established protocols for manipulating the germ line of the mouse, these approaches have opened the way for virtually any form of genetic change to be modeled in the mouse.

2.2 Conventional gene-targeting approaches

The central objective of any targeting experiment is to specifically mutate a chosen sequence, either to inactivate it or to change its function, without altering other sequences within the genome. This is undertaken using germ line competent ES cells, which are first genetically modified and subsequently used to populate the germ line of a chimeric animal and ultimately its progeny. The gene-targeting event requires extremely high-sequence specificity, a condition imposed by reliance upon homologous recombination between endogenous sequences at the target locus and identical or near-identical sequence cloned into the targeting vector of choice.

The first precise changes engineered into the germ line of mice were targeted into the *Hprt* gene (1), primarily because of the ease of selection of inactivated mutant cells. The *Hprt* locus is located on the X chromosome,

Gene Targeting and Embryonic Stem Cells, Alison Thomson and Jim McWhir
© 2004 Garland Science/BIOS Scientific Publishers.

and as the majority of ES cell lines are male, only one targeting event was required to render them *Hprt* deficient and thereby resistant to medium containing 6-thioguanine. The targeting vector designs used in these early experiments were rapidly adopted for a host of other loci, being composed of a number of simple elements. These included homology with the endogenous sequence, a positive selection cassette to allow identification of targeted ES cell clones in culture, and a negative selection cassette to enable enrichment of the targeting events by selecting against random insertion events.

Two basic vector structures were used, either with two arms of homology to allow direct replacement of sequences at the target locus, termed a replacement vector; or with a single region of homology that was normally incorporated in its entirety into the target locus, termed an insertion vector.

Several factors were quickly recognized that modify the efficiency of gene targeting. These included the length of homology included in the vector, and the presence of a double strand break to promote recombination. It also became clear that the efficiency of recombination was greatly increased by use of isogenic DNA in the preparation of the targeting vectors. Thus, te Riele *et al.* (2) reported a 20-fold increase in targeting efficiency at the retinoblastoma locus if the targeting vector was derived from the same genetic background as the ES cells.

Another powerful determinant of successful gene targeting has been the choice of the selection system. Predominantly this has relied upon the inclusion of antibiotic resistance genes as expression cassettes, for example driving the neomycin or puromycin resistance genes; or in the case of the negative selection driving expression of toxic genes such as those encoding the HSV thymidine kinase or diphtheria toxin proteins. Somewhat more efficient gene targeting has been delivered by reliance upon promoter trap strategies, wherein expression of the selection cassette depends upon appropriate insertion downstream of promoter sequences, so enriching for targeted recombinants (3). The primary limitation of this approach is that it cannot be applied to target genes not expressed in ES cells.

The design of gene targeting vectors has not simply been limited to knockout vectors. The need to introduce and study subtle mutations has long been recognized, with many human diseases associated with functional polymorphisms or single point mutations, such as in the delta F508 mutation in cystic fibrosis. The central difficulty in engineering point mutations is that the normal targeting strategy requires a selection cassette to be introduced at the targeted allele. Clearly this is not easily compatible with the creation of single point mutations. Strategies used to circumvent this problem have included simply inserting the selection cassette within either intronic sequences or within 5′ or 3′ UTRs, the use of coelectroporation of a selection cassette (4), and the use of two-stage strategies based either upon insertion vectors (the 'hit and run approach' e.g. 5), or upon replacement vectors, the latter approach referred to as either 'double replacement' (6) or 'tag and replace' (e.g. 7). Perhaps the most widely used of these approaches is the double replacement strategy. Here, conventional replacement vectors are used to inactivate the gene of interest by using flanking homology to delete the area of interest. Critically, this stage is used to introduce both positive and negative selection cassettes into the target locus. A second targeting vector is then used which spans the entire targeted region. This vector contains no selection marker, but carries a mutation of choice. Subsequent recombination and selection against the negative cassette builds the mutated allele and removes both selection cassettes. This approach has proven particularly powerful for the analysis of multiple subtle mutations made within the prion-related protein (PrP) locus (7,8).

2.3   Conditional alleles

The manipulation of endogenous genes using gene-targeting technology is a powerful tool for the analysis of the function of a specific gene product. To date the most common usage of this technology has been in the creation of null mutations. It became rapidly clear that this technology would be significantly enhanced if a given gene could be mutated in a tissue-specific and/or -inducible pattern. This would, for example, allow the study of separate functions of pleiotropic genes or allow analysis of consequences of dysfunction free from any complications arising from constitutive loss. These goals have largely been achieved through use of the Cre-Lox and Flp-Frt systems, which represent essentially equivalent methods to create conditional alleles.

Mutant alleles are generated using a two-step system which utilizes the bacteriophage P1 Cre/LoxP (or equivalent Flp/Frt) site-specific recombination system. Sequences flanked by 34-bp LoxP sites are introduced to the target locus by conventional targeting strategies in such a manner as not to perturb normal expression of the target gene. Sequences flanked by the LoxP or Frt sites can then be deleted *in vivo* or *in vitro* by expression of the Cre or Flp recombinases. *In vivo* this is usually achieved through the generation of a further transgenic line in which the Cre or Flp transgene is expressed in either a tissue-specific or -inducible fashion. The latter has been achieved by fusion to the estrogen response element (e.g. 9), through use of the Tet repressor system (10), and also through use of a p450-inducible promoter (Sansom *et al.*, personal communication). The Cre-Lox and Flp-Frt approaches have become relatively widespread, with a growing list of both conditional alleles and well-characterized Cre/Flp expressing lines (e.g. http://www.mshri.on.ca/nagy/cre.htm).

The Cre-Lox/Flp-Frt systems are also finding widespread use in chromosomal engineering and the manipulation of targeting vectors, both of which are briefly discussed below, and also in creating defined transgenics by mediating integration at a predefined LoxP site. This latter approach relies on the use of two different mutant LoxP sites which fail to regenerate functional LoxP sites post integration and so block re-excision of the transgene (11).

One of the practical drawbacks of the Cre-Lox/Flp-Frt approaches is the time required to generate the appropriate intercrosses once the conditional allele has been passed through the germ line. Thus, in order to analyze an autosomal gene, the conditional allele must be bred to homozygosity on a Cre transgenic background. Seibler *et al.* (12) recently described a significant shortcut, by engineering all the appropriate alleles at the ES cell stage, and then generating experimental mice directly by tetraploid blastocyst injection. This protocol can theoretically reduce the timescale of this type of experiment by 8–10 months, although it is absolutely reliant upon maintaining ES cells through several selection cycles in a state compatible with tetraploid rescue.

2.4 Tracking mutations: coat color tagging

Once targeted ES cells have been generated with a desired mutation, it is of course necessary to introduce this mutation into the germ line and then track that mutation through subsequent generations. The conventional approach to this has been to track the mutation by PCR amplification of DNA extracted from biopsied material. In order to circumvent this rather laborious approach, Zheng *et al.* (13) have used a tyrosinase coat color minigene, introducing it into the mutant locus. Tyrosinase is required for melanin biosynthesis, and therefore its expression leads to pigment production and thereby to a detectable coat color in an albino background that ranges from light gray to gray. This approach can even be used to distinguish between heterozygotes and homozygotes, as has been demonstrated for a targeted p53 allele (13). An agouti minigene has been used in an analogous manner, in this instance expression of the minigene directed by a keratin 14 promotor leading to yellowing of the fur color in a wild-type agouti background (13,14).

2.5 Enhancing gene-targeting efficiency

Considerable effort has been placed into methodologies aimed at enhancing gene-targeting frequencies, both through chemical and genetic intervention. One method known to improve the specificity of gene targeting has been the inhibition of poly(ADP-ribose) polymerase (PARP). Thus, the PARP inhibitor 3-methoxybenzamide has been demonstrated to lower illegitimate recombination, although this does also lower the absolute frequency of gene targeting. Perhaps more promisingly, use of a second PARP inhibitor, 1,5 isoquinolinediol, has been shown to increase by up to 8-fold the absolute gene-targeting frequency at a test allele (15). One notable curiosity concerning the role of PARP in suppressing homologous recombination is that it appears specific to the method of targeting vector delivery. Thus, inhibition of PARP selectively enhances targeting frequencies for calcium phosphate precipitation, but not electroporation (16).

It is very clear that altering the genetic status of the cell undergoing gene targeting is a potent determinant of gene-targeting frequency, perhaps most clearly demonstrated by the avian leukosis virus-induced chicken B cell line DT40 (17), in which targeting frequencies of 10–100% can be readily obtained without the need for specialized selection protocols. The very existence of this line implies that other cell types, including ES cells, should be open to enhancement through genetic manipulation. Perhaps the most obvious candidates for such manipulation are those genes identified as having a role in homologous recombination, DNA damage signaling and repair, and nonhomologous end joining (NHEJ, the process thought to underlie the random integration of targeting vectors).

Several groups have therefore pursued these possibilities. For example, Yanez and Porter (18) investigated the effect of overexpressing hRAD51 in human cells, a protein with homologous DNA pairing and strand exchange activities. These experiments showed a modest (2–3 fold) increase in gene targeting with no obvious detrimental cellular effects. Recently, Dominguez-Bendala and colleagues have shown a very similar effect of Rad51 overexpression within ES cells (19). Overexpression of different repair mechanisms appears to have a variable impact upon targeting efficiency. Thus, overexpression of Rad52 suppresses gene targeting whilst stimulating extrachromosomal homologous recombination, possibly through suppression of Rad51 activities (18).

Mutation of other components of the homologous recombination/repair machinery can also reduce targeting efficiencies. Thus deficiency of one of the two Rad54 homologs in ES cells has been shown to markedly reduce targeting efficiency, probably as a consequence of loss of its normal role in stimulating Rad51-mediated strand exchange (20).

The tumor suppressor proteins BRCA1 and BRCA2 both interact with Rad51, and deficiency of Brca1 in ES cells reduces the frequency of both intrachromosomal homologous recombination and gene targeting, whilst at the same time increasing the frequency of random integration (21,22). The possibility that Brca1 overexpression may enhance gene-targeting frequencies has yet to be explored.

The consequences of p53 deficiency have also been investigated, with reported rises in extrachromosomal and intrachromosomal homologous recombination of 10–100 fold (23,24), which argue strongly that transient inactivation of p53 may well stimulate gene targeting. These data however contrast with reports of normal levels of homologous recombination in ES cells deficient for p53 (19). These apparent differences in p53 dependency may well reflect the availability of p53 within ES cells, as p53 appears to be at least partially inactive in ES cells due to both cytoplasmic sequestration of the protein and to downstream deficiencies in the p53-mediated response (25).

Deficiency of the DNA mismatch repair genes has also been shown to alter gene-targeting efficiencies, by reducing the requirement for absolute homology within the targeting vector. Thus, mismatch repair strongly suppresses homologous recombination when the construct and target locus diverge at <1% of nucleotide positions. More recently, mismatch repair has also been shown to suppress gene modification mediated by small synthetic DNA oligonucleotide sequences (26).

With respect to NHEJ, it remains somewhat unclear whether manipulating the genes that have been implicated in this pathway, such as Ku70, Ku80, Xrcc4 and ligase IV, will be valuable in ES cells. However this area remains of great interest, given that the enhanced targeting efficiency of the DT40 line is essentially derived from a reduction in random integration, presumably through perturbed NHEJ.

One of the earliest observations was that gene targeting is enhanced by the presence of double-strand breaks, raising the possibility that the introduction of double-strand breaks may be used to enhance targeting frequencies. Both restriction enzymes and rare cutters such as I-*SceI* have indeed been used in this fashion in a range of cell types, including ES cells (27). The limitation of this approach in ES cells is that the double-strand break must be selectively introduced at the endogenous target gene sequences, which is effectively impossible for frequent cutters, and requires prior targeting for the rare cutters such as I-*SceI* in order to introduce the restriction site at the desired location. Nonetheless, this approach has clear application in the repeated targeting of a given allele, for example to facilitate the introduction of a series of point mutations.

It is also possible that systems developed in other organisms will soon become available for mammalian targeting experiments. For example, Bibikova *et al*. (28) have recently shown that zinc finger nucleases can be used to stimulate gene targeting in Drosophila, essentially by introducing double-strand breaks into the target locus. Although this has so far only been demonstrated for a given locus within Drosophila, this approach may be broadly applicable as the ability of double-strand breaks to stimulate recombination is a property of essentially all cells and organisms, and indeed is already recognized as a critical factor in conventional gene targeting through the linearization of vectors.

The drive to utilize homologous recombination within human gene therapy treatment regimes has led to the development of a series of adeno-associated viral (AAV) vectors which package gene-targeting vectors as single-stranded linear molecules. Such AAV-mediated gene targeting has been shown to be significantly more efficient than conventional targeting strategies, although still as yet below clinically useful thresholds. The mechanism by which AAV augments gene targeting remains unclear, although it is presumed to lie either in the unique topology of the AAV vector genome, or in the efficient nuclear delivery of single-stranded targeting constructs. Notably, AAV-mediated gene targeting has recently been shown to be significantly enhanced by double-strand breaks (29,30).

2.6 Targeted mutagenesis using triplex-forming oligonucleotides

Triple-helix-forming oligonucleotides (TFOs) recognize and bind specific sequences via the major groove of duplex DNA, and may offer an alternative mechanism for precisely engineering the genome. Their potential for targeted mutagenesis derives from the ability of TFOs to precisely deliver a chemical damaging agent (psoralen), which then mediates site-specific mutagenesis. TFOs bind to unique sites which occur approximately once per kilobase, with hydrogen bonding between the TFO and the purine-rich strand of the target DNA. The psoralen linked TFO can subsequently be activated following exposure to UVA light, which results in interstrand crosslinking and subsequent repair; a process which incidentally stimulates homologous recombination. Significantly, the outcome of this repair is frequent conversion into mutations, leaving the endogenous target altered (e.g. 31). The critical parameters governing TFO-mediated mutation remain to be determined, as does the feasibility of this approach in ES cells; however this remains a potentially exciting addition to the gene-targeting armory.

2.7   Chromosome engineering

Gene-targeting strategies have allowed the engineering of individual loci, however the major genetic mechanism underlying human disease is the rearrangement of chromosomes. Targeted breakage of chromosomes had been achieved as early as 1992 by Itzhaki *et al*. (32) who used gene targeting of telomeric DNA to initiate a defined truncation. However, a more generalized method first became available in 1995, through utilization of a Cre-Lox-based approach. Ramirez-Solis *et al*. (33) showed that two targeting events could be used to introduce Lox P sites at extremely distant sites. The targeting was designed such that following Cre-mediated recombination, a positive selection marker was regenerated, so allowing selection of recombinant ES cell clones. Practically this was achieved by incorporating complementary but nonfunctional fragments of *Hprt* at each targeted locus, and then driving recombination by transient expression of a Cre-expressing vector to restore *Hprt* function. The chromosomal variant generated can either be a duplication, deletion or inversion of the chromosomal segment dependent upon the orientation of the LoxP sites. This approach can also be exploited to deliver better genetic screens in the mouse. In other organisms, such as *Drosophila*, chromosomal variants are commonly exploited as a small portion of the genome is functionally hemizygous. This is not the case in the mouse, rendering the genetic screen for recessive mutations very cumbersome. The ability to create defined chromosomal deficiencies, inversions and duplications effectively creates mice which show segmental haploidy, and which can therefore be used in genetic screens.

2.8   High-throughput gene targeting strategies

2.8.1   High-throughput vector construction

Although targeting technology has led to the development of a plethora of new murine mutant strains, the conventional strategies remain somewhat laborious. Whilst this has not proven a barrier to single gene analysis, it has made high-throughput approaches difficult. One solution to this problem has been the generation of genomic libraries in which the genetic elements for gene targeting and coat color tagging are inherent in the vector backbone. To this end, Zheng *et al*. (34) created two phage genomic libraries, termed 5'hprt and 3'hprt. The former of these contains a neomycin cassette, a tyrosinase minigene, a plasmid backbone and origin of replication, a random genomic insert and the 5' half of the *Hprt* minigene. These elements are all flanked by two LoxP sites and finally the lambda arms. The 3'hprt library differs in carrying the agouti coat color transgene and the complementary 3'hprt sequences. Both of the libraries were constructed with automatic plasmid excision capabilities, such that following introduction into a Cre-expressing bacterial strain, a fully functional targeting plasmid is produced following Cre-mediated recombination of the lox P sites. The average insert size for these two libraries is 7.6 and 9.1 kb, a size adequately suited to insertional gene targeting. In addition to their use for high-throughput targeting of single genes, these libraries can be used together to generate large chromosomal deletions. ES cells are sequentially targeted with vectors derived respectively from the 5′ and 3′ libraries that contain homology to the two deletion endpoints. These are then exposed to Cre recombinase activity, and selected in HAT medium for HPRT expression. Depending upon the orientation and location of loxP sites, this general scheme can be used to generate deletions, duplications, inversions and translocations.

An absolute requirement of the gene-targeting process is that the targeting event be confirmed using a probe external to the homology. The importance of the external nature of the probe is that such probes cannot hybridize with random integrations of the targeting construct to generate events that mimic the recombinant locus. A potential drawback to the targeted library approach is that such a flanking probe is unavailable. However, this problem can be circumvented by relying upon gap repair insertional gene targeting. In this scenario, a gap is introduced into the homologous sequence in the targeting vector, which is repaired from the chromosomal strand during targeted recombination. The 'gap' sequence then serves as an external probe to the targeting event that cannot hybridize to random integrants. This 'gapped' targeting strategy has been successfully used at a number of different loci with targeting frequencies ranging from 4–64% obtained, and appears to offer a good methodology for high-throughput, coat color tagged insertional gene targeting.

As discussed above, not only should such an approach facilitate the rapid generation of individual targeting events, but it also allows the generation of vectors for chromosomal engineering. Indeed, the two lines detailed above were specifically developed for this, as they bear independent selection cassettes, independent coat color markers and complementary nonfunctional Hprt sequences. The ability to facilitate high-throughput chromosomal modifications may become very relevant to the development of phenotype-driven mutagenesis screens, an approach which has proved invaluable in screens for recessive mutations in Drosophila (35). Essentially this approach relies on segmental haploidy to recover recessive mutations. Such haploidy can now be delivered in the mouse through defined chromosomal engineering.

The 3′hprt and 5′hprt libraries are but two of a number of systems developed to deliver high-throughput gene targeting. A somewhat similar approach has been adopted by Wattler *et al.* (36), who used a lambda phage murine 129 SV/Ev^{brd} genomic library which contains negative selection cassettes flanking the genomic inserts. The library is then screened by PCR to identify a lambda clone appropriate to the desired locus to be targeted, which is then converted to a high copy plasmid using a parallel Cre-LoxP-based self-excision process to that described above for the 3′/5′hprt libraries. A yeast selectable marker is then tagged with 30–40 base pairs of homology to the target site and cotransformed into yeast with the genomic clone, which permits introduction of the yeast selectable marker into the homology. A one-directional cloning step is then used to insert a custom-targeting cassette (for example the introduction of a positive/negative selection cassette) using the restriction sites introduced during the yeast recombination step. This approach has the advantage that the end product is a replacement-type vector, and furthermore that the final structure of the targeting cassette can be predetermined by relevant choices made during the lambda screen and the yeast recombination.

Similar strategies have used bacterial recombination systems to effectively rescue target genomic sequences into targeting vectors by using linear cloning vectors containing a bacterial origin and selection cassette, as well as being flanked by short (>70 base pair) sequences of homology to the target locus (37). Cotransformation of genomic DNA with these vectors results in the rescue of the desired genomic sequence into the targeting vector. Although these methods do not provide an immediate 'bank' of targeting vectors, they have provided strategies which allow very rapid and efficient targeting vector production.

2.8.2 High-throughput gene trapping

An alternative approach to high-throughput mutation of the murine germ line has been through the creation of large gene-trapped libraries. Gene-trapping vectors essentially emerged out of the efforts to develop promoter trap-targeting vectors as described above, and they are designed such that they only become transcriptionally active when they are inserted into an endogenous gene. This is typically achieved by creating a vector that lacks an enhancer, promoter or polyadenylation signal, or indeed any combination of these. Gene trapping is essentially designed to inactivate loci at random, and so has great potential for high-throughput strategies. Perhaps the most extensive library to date has been developed commercially by Lexicon genetics (http://www.lexgen.com/omnibank/overview.php), who report more than 200 000 knockout mouse embryonic stem cell clones which is purported to correspond to greater than 50% of mammalian genes. The strategy used to generate these clones was to utilize gene trap vectors based upon two functional units: first, a sequence acquisition component consisting of the phosphoglycerate kinase-1 promoter fused to a puromycin N-acetylase gene which lacked a polyadenylation sequence but which was followed by a synthetic splice donor sequence; second, a mutagenic component consisting of a splice acceptor sequence fused to a selectable colorimetric marker gene followed by a polyadenylation sequence. By combining these elements gene trap vectors were generated with a demonstrated high ability to mutagenize and which could also trap genes independent of transcription status. Coupled together with high-throughput identification of the target loci using 3'RACE, this approach has succeeded in delivering an extensive library of trapped genes, which should complement the targeting-based strategies already described.

2.8.3 Gene targeting with bacterial artificial chromosomes (BACs)

One of the earliest identified determinants of targeting efficiency was the length of homology used in the targeting vector, with higher targeting efficiencies correlating with greater homology. The precise reasons for this remain somewhat unclear as, somewhat surprisingly, the search for homology is not a rate-limiting step in mammalian cells (38). An alternative explanation for the effect of increased homology is that favored sites exist for the initiation of DNA exchange, and that these may be considerably distant to each other. One logical approach is therefore to dramatically increase the region of homology using BAC-based targeting vectors. This has recently been achieved by Yang and Seed (39), who utilized bacterial recombination systems similar to those discussed above to generate a series of BAC-based targeting vectors. Targeted ES cell clones were generated at five different loci using this approach and subsequently confirmed using both PCR and FISH. The mean ratio of homologous to nonhomologous recombination events was 1.5×10^{-1}, which compares favorably with conventional approaches (which range between 10^{-2} and 10^{-5}). A similar scientific approach has been adopted by Valenzuela *et al.* (40), but in a high-throughput and largely automated context (termed 'VelociGene') that allowed the rapid production of 200 individual targeted lines. Apart from establishing the feasibility of high-throughput BAC-based targeting, the data generated in this study indicate that BAC-based targeting may alleviate some of the constraints upon targeting design, such as the requirement for isogenic sequences. Valenzuela *et al.* addressed the problem of confirming the targeting event by using real time quantitative PCR directed at the deletion event to be introduced. Thus, ES clones undergoing the correct targeting event were identified by virtue of haploinsufficiency at the region of deletion. As with the Yang and Seed study (39), these events were subsequently confirmed using FISH.

2.9 Summary

The ability to manipulate the murine genome has proven to be one of the most powerful forces driving modern biology. The first gene-targeted strain was developed 15 years ago, and since then multiple strategies have been developed which permit the generation of virtually any desired genetic modification. The most recent advances in the field have focused upon promoting the efficiency of gene targeting, predominantly through manipulating those genes implicated in homologous recombination and DNA repair, and upon the creation of high-throughput gene-targeting protocols. Indeed, given the availability of the entire genome sequence and the consequent plethora of potential questions to be resolved, it is advances in this latter area that are perhaps the most crucial to how we tackle the biological challenges of the post genomic era.

References

1. Thompson S, Clarke AR, Pow AM, Hooper ML, Melton DW (1989) Germ line transmission and expression of a corrected HPRT gene produced by gene targeting in embryonic stem cells. *Cell* **56**: 313–321.
2. Te Riele H, Maandag ER, Berns A (1992) Highly efficient gene targeting in embryonic stem cells through homologous recombination with isogenic DNA constructs. *Proc Natl Acad Sci USA* **89**: 5128–5132.
3. Sedivy JM, Dutriaux A (1999) Gene targeting and somatic cell genetics – a rebirth or a coming of age? *Trends Genet* **15**: 88–90.
4. Davis AC, Wims M, Bradley A (1992) Investigation of coelectroporation as a method for introducing small mutations into embryonic stem cells. *Mol Cell Biol* **12**: 2769–2776.
5. Lakhlani PP, MacMillan LB, Guo TZ, McCool BA, Lovinger DM, Maze M, Limbird LE (1997) Substitution of a mutant alpha2a-adrenergic receptor via "hit and run" gene targeting reveals the role of this subtype in sedative, analgesic, and anesthetic-sparing responses in vivo. *Proc Natl Acad Sci USA* **94**: 9950–9955.
6. Stacey A, Schnieke A, McWhir J, Cooper J, Colman A, Melton DW (1994) Double-replacement gene targeting to replace the murine alpha-lactalbumin gene with its human counterpart in embryonic stem-cells and mice. *Mol Cell Biol* **14** (2): 1009–1016.
7. Moore RC, Redhead NJ, Selfridge J, Hope J, Manson JC, Melton DW (1995) Double replacement gene targeting for the production of a series of mouse strains with different prion protein gene alterations. *Biotechnology* **13**: 999–1004.
8. Barron RM, Manson JC (2003) A gene-targeted mouse model of P102L Gerstmann-Straussler-Scheinker syndrome. *Clin Lab Med* **23**: 161–173.
9. Forde A, Constien R, Grone HJ, Hammerling G, Arnold B (2002) Temporal cre-mediated recombination exclusively in endothelial cells using the Tie2 regulatory elements. *Genesis* **33**: 191–197.
10. Schonig K, Schwenk F, Rajewsky K, Bujard H (2002) Stringent doxycycline dependent control of CRE recombinase in vivo. *Nucleic Acids Res* **30**: e134.
11. Araki K, Araki M, Yamamura K (2002) Site directed integration of the cre gene mediated by Cre recombinase using a combination of mutant lox sites. *Nucl Acids Res* **30**: e103.
12. Seibler J, Zevnik B, Kuter-Luks B, *et al.* (2003) Rapid generation of inducible mouse mutants. *Nucl Acids Res* **31**: 4e12.
13. Zheng B, Vogel H, Donehower LA, Bradley A (2002) Visual genotyping of a coat color tagged p53 mutant mouse line. *Cancer Biol Ther* **1**: 433–435.
14. Kucera GT, Bortner DM, Rosenberg MP (1996) Overexpression of an Agouti cDNA in the skin of transgenic mice recapitulates dominant coat color phenotypes of spontaneous mutants. *Dev Biol* **173**: 162–173.
15. Semionov A, Cournoyer D, Chow TY (2003) 1,5-isoquinolinediol increases the frequency of gene targeting by homologous recombination in mouse fibroblasts. *Biochem Cell Biol* **81**: 17–24.
16. Stricklett PK, Nelson RD, Kohan DE (1999) The Cre/loxP system and gene targeting in the kidney. *Am J Physiol* **276**: F651–F657.
17. Buerstedde JM, Takeda S (1991) Increased ratio of targeted to random integration after transfection of chicken B cell lines. *Cell* **67**: 179–188.
18. Yanez RJ, Porter AC (2002) Differential effects of Rad52p overexpression on gene targeting and extrachromosomal recombination in a human cell line. *Nucl Acids Res* **30**: 740–748.
19. Dominguez-Bendala J, Priddle H, Clarke A, McWhir J (2003) Elevated expression of exogenous Rad51 leads to identical increases in gene-targeting frequency in murine embryonic stem (ES) cells with both functional and dysfunctional p53 genes. *Exp Cell Res* **286**: 298–307.

20. Essers J, Hendriks RW, Swagemakers SMA, *et al.* (1997) Disruption of mouse RAD54 reduces ionizing radiation resistance and homologous recombination. *Cell* **89**: 195–204.
21. Snouwaert JN, Gowen LC, Latour AM, Mohn AR, Xiao A, DiBiase L, Koller BH (1999) BRCA1 deficient embryonic stem cells display a decreased homologous recombination frequency and an increased frequency of non-homologous recombination that is corrected by expression of a brca1 transgene. *Oncogene* **18**: 7900–7907.
22. Moynahan ME, Chiu JW, Koller BH, Jasin M (1999) BRCA2 is required for homology-directed repair of chromosomal breaks. *Mol Cell* **4**: 511–518.
23. Gebow D, Miselis N, Lieber HL (2000) Homologous and nonhomologous recombination resulting in deletion: effects of p53 status, microhomology, and repetitive DNA length and orientation. *Mol Cell Biol* **20**: 4028–4035.
24. Mekeel KL, Tang W, Kachnic LA, Luo CM, DeFrank JS Powell SN (1997) Inactivation of p53 results in high rates of homologous recombination. *Oncogene* **14**: 1117–1122.
25. Aladjem MI, Spike BT, Rodewald LW, Hope TJ, Klemm M, Jaenisch R, Wahl GM (1998) ES cells do not activate p53-dependent stress responses and undergo 53-independent apoptosis in response to DNA damage. *Curr Biol* **8**: 145–155.
26. Dekker M, Brouwers C, te Riele H (2003) Targeted gene modification in mismatch-repair deficient embryonic stem cells by single strand DNA oligonucleotides. *Nucl Acids Res* **31**: e27.
27. Cohen-Tannoudji M, Robine S, Choulika A, *et al.* (1998) I-Sce induced gene replacement at a natural locus in embryonic stem cells. *Mol Cell Biol* **18**: 1444–1448.
28. Bibikova M, Beumer K, Trautman JK, Carroll D (2003) Enhancing gene targeting with designed zinc finger nucleases. *Science* **300**: 764.
29. Porteous MH, Cathomen T, Weitzman MD, Baltimore D (2003) Efficient gene targeting mediated by adeno associated virus and double-strand breaks. *Mol Cell Biol* **23**: 3558–3565.
30. Miller DG, Petek LM, Russell DW (2003) Human gene targeting by adeno-associated virus vectors is enhanced by DNA double-strand breaks. *Mol Cell Biol* **23**: 3550–3557.
31. Majumar A, Puri N, Cuenoud B, *et al.* (2003) Cell cycle modulation of gene targeting by a tri helix-forming oligonucleotide. *J Biol Chem* **278**: 11072–11077.
32. Itzhaki JE, Barnett MA, MacCarthy AB, Buckle VJ, Bro WR, Porter AC (1992) Targeted breakage of a human chromosome mediated by cloned human telomeric DNA. *Nature Genet* **2**: 283–287.
33. Ramirez-Solis R, Liu P, Bradley A. (1995) Chromosomal engineering in mice. *Nature* **378**: 720–724.
34. Zheng B, Mills AA, Bradley A (1999) A system for rapid generation of coat colour tagged knockouts and defined chromosomal rearrangements in mice. *Nucl Acids Res* **27**: 2354–2360.
35. Roberts DB (ed) (1998) *Drosophila: A practical approach*. IRL Press, Oxford, UK.
36. Wattler S, Kelly M, Nehls M (1999) Construction of gene targeting vectors from lambda KOS genomic libraries. *Biotechniques* **26**: 1150–1159.
37. Angrand PO, Daigle N, van der Hoeven F, Scholer HR, Stewart AF (1999) Simplified generation of targeting vectors using ET recombination. *Nucl Acids Res* **27**: e16.
38. Zheng H, Wilson JH (1990) Gene targeting in normal and amplified cell lines. *Nature* **344**: 170–173.
39. Yang Y, Seed B (2003) Site specific gene targeting in mouse embryonic stem cells with intact bacterial artificial chromosomes. *Nature Biotechnol* **21**: 447–451.
40. Valenzuela DM, Murphy AJ, Frendewey D, *et al.* (2003) High throughput engineering of the mouse genome coupled with high resolution expression analysis. *Nature Biotechnol* **21**: 652–659.

Nuclear transfer with murine embryonic stem cells

3

Shaorong Gao

3.1 Introduction

In 1996, a landmark breakthrough was made by Wilmut and his colleagues. A sheep named Dolly was successfully cloned from an adult somatic cell by electro-fusion of a nuclear donor cell from the mammary gland to an enucleated metaphase II oocyte (1). This advance demonstrated that cellular commitment in the adult was not irreversible and has led to the further development of this important technology.

3.1.1 Cloning by nuclear injection

In 1998, Wakayama *et al.* (2) used direct nuclear injection to clone mice from cumulus cells. A Piezo-drill micromanipulator was employed to allow direct injection of nuclei into enucleated metaphase II (MII) oocytes. After activation, reconstructed embryos were allowed to develop to morulae/blastocyst stage before transfer into surrogate mothers. The piezo-drill micromanipulation technique was first established by Kimura in 1995 (3,4) to generate mice by intracytoplasmic sperm injection (ICSI). Although at one time low temperatures were necessary during nuclear injection, with improvements to Piezo-drill equipment and manipulation techniques, room temperature is now widely used. Compared to the electrofusion method used for large animal cloning, piezo-drill microinjection leads to less donor cell cytoplasm being introduced into the reconstructed ocyte.

3.1.2 Modeling, reprograming and therapeutic cloning in mice

Unlike large animals, mice have a shorter gestation and generation period and more information is known about the genetic background of mice. So it is possible to use the mouse as a model to study the basic mechanism of reprograming following nuclear transfer. Cloning of mice also provides an alternate and more efficient means to produce transgenic or knockout mice. To date, cloned mice have been successfully produced from both somatic cells and embryonic stem (ES) cells irrespective of hybrid or inbred background (5–11). ES cells are cell lines derived from the inner cell mass of blastocysts, isolated either from normal fertilized embryos or cloned blastocysts derived from somatic cells (12). ES cells are karyotypically and

Gene Targeting and Embryonic Stem Cells, Alison Thomson and Jim McWhir
© 2004 Garland Science/BIOS Scientific Publishers.

phenotypically stable after long-term culture *in vitro*, allowing us to make genetic modifications to these cells before nuclear transfer.

Because ES cells can be differentiated into several types of cells *in vitro*, therapeutic cloning provides a means by which therapeutically useful cells can be derived from specific genotypes. Since the technology is more advanced in mice and oocytes more readily available, mice have been studied as a model of human therapeutic cloning (13). Murine ES cells were successfully derived from somatic cell-cloned blastocysts and further differentiated *in vitro* to produce neural and bone cells. These differentiated cells were transplanted back to mice defective in these tissues, to observe if the transplanted cells can survive and help the host organ to be rescued. Ultimately we would hope to use this technique to restore organ function or to prevent or slow further deterioration in humans. However, attempts to clone human embryos have so far proven unsuccessful, possibly because of the loss of critical factors in primates during enucleation (14).

Since mice have been successfully cloned from both somatic and ES cells by Wakayama, many laboratories in the world have tried to repeat his achievement. However, the required micromanipulation skills are demanding and so far only a few labs can make cloned mice by Piezo-drill micromanipulation. Other important factors include detailed aspects of the preparation of both solutions and cells. In our recent studies, we have successfully produced cloned mice from both hybrid and inbred ES cells and cumulus cells by using the so-called Honolulu method (11). In this chapter, the methods of preparation of solutions, donor cells and nuclear transfer techniques are described in detail. We anticipate that in time many more laboratories will master this technique to explore the basic mechanism of reprograming and to further develop therapeutic cloning for human therapy.

3.2 Factors affecting efficiency of nuclear transfer

3.2.1 Strain of origin of ES cells affects nuclear transfer

Mouse embryonic stem (ES) cells are derived from the inner cell mass of blastocysts. Compared to somatic cells, they exhibit karyotypic and phenotypic stability with long-term culture and they are more suitable for *in vitro* modification to make transgenic cloned mice. Hybrid and inbred ES cells were compared as nuclear donors in nuclear transfer studies to determine the effect of genotype on developmental potential of cloned embryos (9). The results showed that most hybrid ES cell clones survived while all inbred ES cell cloned mice died, suggesting that heterozygosity of donor ES cells is essential for survival of cloned mice (9). In contrast to this study, we have successfully cloned mice from both inbred and hybrid ES cells, although the survival rate of inbred clones was slightly lower (11). We consider that both the mouse strain from which the ES cell is derived and the method used for culture and preparation of ES cells are important for the survival of cloned mice. The inbred ES cell line we successfully used is HM-1 (15), which is derived from 129/Ola. After transfer of cloned embryos at morulae/blastocyst stage to pseudo-pregnant mice, 18 pups were successfully cloned using HM-1 ES cells at passage 19 and five cloned mice survived to adulthood (*Table 3.1*).

Table 3.1 Effect of ES cell confluence on development of NT embryos reconstructed with inbred HM-1 ES cells

Confluence (%) of ES cells	No. of enucleated oocytes injected	No. (%) of oocytes survived injection	No. (%) of activated oocytes	No. (%) of cleaved oocytes	No. (%) of morulae/blastocysts developed	No. of transferred embryos (recipients)	No. (%) of dead fetuses	No. (%) of live fetuses
60–70	205	196 (95.6)	194 (99.0)	115 (59.3)[a]	43 (22.2)[a]	43 (3)	0	1 (2.3)[a]
80–90	436	420 (96.3)	412 (98.1)	349 (84.7)[b]	202 (49.0)[b]	197 (12)	9 (4.6)	18 (9.1)[b]

[a], [b] Significant difference ($P < 0.01$).

Table 3.2 *In vitro* and *in vivo* development of cloned embryos reconstructed with ED1 and ED2 hybrid ES cells

ES cell used for nuclear transfer	No. of enucleated oocytes injected	No. (%) of oocytes survived injection	No. (%) of activated oocytes	No. (%) of cleaved oocytes	No. (%) of morulae/blastocysts developed	No. of transferred embryos (recipients)	No. (%) of dead fetuses	No. (%) of live fetuses
ED2	330	317 (96.1)	310 (97.8)	258 (83.2)	138 (44.5)	138 (13)	2 (1.4)	19 (13.8)
ED1	267	256 (95.9)	249 (97.3)	205 (82.3)	100 (40.2)	100 (9)	0	0

[a], [b] Significant difference ($P < 0.01$).

These five mice have proven fertile. To investigate the developmental potential of cloned embryos reconstructed with hybrid ES cells, we isolated two F1 (C57BL/6J × 129/SV) ES cell lines (ED1 and ED2) and both carried the normal male karyotype. After nuclear transfer, reconstructed embryos at morulae/blastocyst stage were transferred into recipients. As shown in *Table 3.2*, 13.9% of transferred embryos reconstructed with ED2 ES cells developed into live pups and 11 of 19 (60%) survived to adulthood. This result agrees with previous results and the survival rate of ED2 hybrid ES cell clones was slightly higher than inbred ES cell clones in our study. However, no pups were produced from ED1 reconstructed embryos. Our results show that the inbred ES cell line HM-1 can be used to produce adult clones and that not all hybrid ES lines produce cloned mice.

3.2.2 Confluence and passage number affect the properties of nuclear donor cells

ES cell confluence during culture plays an important role in the successful development of cloned embryos. As shown in *Table 3.1*, when HM-1 ES cells at passage 19 were used for nuclear transfer studies, cells taken at low confluence were less effective nuclear donors than cells taken at high confluence as measured by both pre- and postimplantation development. No pups cloned from cells at low confluence survived to adulthood, while five of 18 cloned pups from high-confluence cells developed into normal fertile mice. It might be that expression of early developmental genes is more appropriate after nuclear transfer from high-confluence cells, though why this should be the case is unclear.

Cell passage number also affects the developmental potential of cloned embryos. Both our own data (11) and that of others (16) demonstrate that no cloned embryos reconstructed with R1 ES cells over passage 20 could develop into live pups. With increasing cell passage number, the proportion of cells with abnormal karyotype increases and this may account for poor development of cloned embryos.

Cell cycle coordination of donor cells with recipient oocytes has been considered a crucial factor for successful development of cloned embryos (17). Synchronized and unsynchronized ED2 nuclear donor cells were compared for normal development following nuclear transfer. The results showed that even preimplantation development of M-phase ES cell reconstructed embryos was much higher than embryos reconstructed with unsynchronized (presumably G1) cells, however, term development was much lower and most of the cloned pups died after C-section. Damage to condensed metaphase chromosomes during micromanipulation might lead to the low term development and low survival rate.

3.2.3 Origin of donor oocytes

Another factor affecting development of reconstructed embryos is the origin of the donor oocyte. In a previous study, no embryos reconstructed with inbred oocytes and cumulus cells could develop to term while B6D2F1 oocytes could support full-term development of embryos reconstructed with both somatic and ES cells. Heterozygosity of donor oocytes may be crucial for development of cloned embryos (6). We compared developmental potential of reconstructed embryos using two different F1 oocytes: B6D2F1 (C57BL/6 × DBA2) and B6CBAF1 (C57BL/6 × CBA). The results showed that both types of oocytes could support full-term development of embryos reconstructed with ES cells. B6D2F1 is preferred to B6CBAF1 for supplying donor oocytes because it is much easier in that strain to visualize the MII spindle area, making enucleation relatively easy.

3.2.4 Culture conditions

CZB medium is widely used for culturing cloned embryos reconstructed with various donor cells. Several studies have investigated the influence of culture medium on development of nuclear transfer embryos reconstructed with cumulus cells (18,19). The results suggested that a combination of CZB medium with KSOM or Witten's medium produced best development of cloned embryos. We investigated the effects of different culture media on development of embryos reconstructed with ES cells and found that M16 medium is superior to CZB medium. We compared the total cell number of reconstructed embryos cultured to blastocysts in these two media and found higher cell number in blastocysts cultured in M16 medium even though blastocyst development showed no difference. The term development of embryos cultured in M16 was also slightly higher than in CZB medium.

Oxygen concentration during culture is another potential factor affecting development of cloned embryos. However, 5% oxygen during culture showed no beneficial effect on development over embryos cultured in 20% oxygen (20). Total cell number in cloned blastocysts cultured under different oxygen concentrations showed no difference, but no embryos developed into fetuses when 5% oxygen was employed. In contrast, term development has been achieved by culturing cloned embryos with 20% oxygen.

Acknowledgments

The author would like to thank Mrs Michelle McGarry, Tricia Ferrier, Judy Fletcher and Drs Ed Gallagher, Jim McWhir and Ian Wilmut from Roslin Institute for their kind help in ES cell derivation, culture and experimental design. The author also appreciates Dr Kevin Eggan from Dr Rudolf Jaenisch's lab for his kind communication on piezo micromanipulation.

References

1. Wilmut I, Schnieke AE, McWhir J, Kind AJ, Campbell KHS (1997) Viable offspring derived from fetal and adult mammalian cell. *Nature* **285**: 810–813.
2. Wakayama T, Perry ACF, Zuccotti M, Johnson KR, Yanagimachi R (1998) Full-term development of mice from enucleated oocytes injected with cumulus cell nuclei. *Nature* **394**: 369–374.
3. Kimura Y, Yanagimachi R (1995) Introcytoplasmic sperm injection in the mouse. *Biol Reprod* **52**: 709–720.
4. Kimura Y, Yanagimachi R (1995) Mouse oocytes injected with testicular spermatozoa or round spermatids can develop into normal offspring. *Development* **121**: 2397–2405.
5. Wakayama T, Yanagimachi R (1999) Cloning of male mice from adult tail-tip cells. *Nat Genet* **22**: 127–128.
6. Wakayama T, Yanagimachi R (2001) Mouse cloning with nucleus donor cells of different age and type. *Mol Reprod Dev* **58**: 376–383.
7. Ogura A, Inoue K, Ogonuki N, *et al.* (2000) Production of male cloned mice from fresh, cultured, and cropreserved immature sertoli cells. *Biol Reprod* **62**: 1579–1584.
8. Wakayama T, Rodriguez I, Perry ACF, Yanagimachi R, Mombaerts P (1999) Mice cloned from embryonic stem cells. *Proc Natl Acad Sci USA* **96**: 14984–14989.
9. Eggan K, Akutsu H, Loring J, *et al.* (2001) Hybrid vigor, fetal overgrowth, and viability of mice derived by nuclear cloning and tetraploid embryo complementation. *Proc Natl Acad Sci USA* **98**: 6209–6214.
10. Rideout III WM, Wakayama T, Wutz A, *et al.* (2000) Generation of mice from wild-type and targeted ES cells by nuclear cloning. *Nat Genet* **24**: 109–110.
11. Gao S, McGarry M, Ferrier T, *et al.* (2003) Effect of cell confluence on production of cloned mice using an inbred embryonic stem cell line. *Biol Reprod* **68**: 595–603.
12. Wakayama T, Tabar V, Rodriguz I, Perry ACF, Studer L, Mombaets P (2001) Differentiation of embryonic stem cell lines generated from adult somatic cells by nuclear transfer. *Science* **292**: 740–743.
13. Rideout III WM, Hochedlinger K, Kyba M, Daley GQ, Jaenisch R (2002) Correction of a genetic defect by nuclear transplantation and combined cell and gene therapy. *Cell* **109**: 17–27.
14. Simerly C, Dominko T, Navara C, *et al.* (2003) Molecular correlates of primate nuclear transfer failures. *Science* **300**: 297.
15. Magin TM, McWhir J, Melton DW (1992) A new mouse embryonic stem cell line with good germ line contribution and gene targeting frequency. *Nucleic Acids Res* **20**: 3795–3796.
16. Zhou Q, Jouneau A, Brochard V, Adenot P, Renard JP (2001) Developmental potential of mouse embryos reconstructed from metaphase embryonic stem cell nuclei. *Biol Reprod* **65**: 412–419.
17. Campbell KH, Ritchie WA, Wilmut I (1993) Nuclear–cytoplasmic interactions during the first cell cycle of nuclear transfer reconstructed bovine embryos: implication for deoxyribonucleic acid replication and development. *Biol Reprod* **49**: 933–942.
18. Chung YG, Mann MR, Bartolomei MS, Latham KE (2002) Nuclear–cytoplasmic "tug of war" during cloning: effects of somatic cell nuclei on culture medium preferences of preimplantation cloned mouse embryos. *Biol Reprod* **66**: 1178–1184.
19. Heindryckx B, Rybouchkin A, Van Der Elst J, Dhont M (2001) Effect of culture media on in vitro development of cloned mouse embryos. *Cloning* **3**: 41–50.

20. Gao S, McGarry M, Priddle H, *et al.* (2003) Effect of donor oocytes and culture conditions on development of cloned mice embryos. *Mol Reprod Dev* **66**: 126–133.
21. Chatot CL, Ziomet A, Bavister BD, Lewis JL, Torres I (1989) An improved culture medium supports development of random-bred 1-cell mouse embryos in vitro. *J Reprod Fertil* **86**: 679–688.
22. Wakayama T, Yanagimachi R (2001) Effect of cytokinesis inhibitors, DMSO and the timing of oocyte activation on mouse cloning using cumulus cell nuclei. *Reproduction* **122**: 49–60.
23. Ogura A, Inoue K, Takano K, Wakayama T, Yanagimachi R (2000) Birth of mice after nuclear transfer by electrofusion using tail tip cells. *Mol Reprod Dev* **57**: 55–59.
24. Ono Y, Shimozawa N, Muguruma K, *et al.* (2001) Production of cloned mice from embronic stem cells arrested at metaphase. *Reproduction* **122**: 731–736.
25. Amano T, Kato Y, Tsunoda Y (2001) Full-term development of enucleated mouse oocytes fused with embryonic stem cells from different cell lines. *Reproduction* **121**: 729–733.

Protocols

Contents

EQUIPMENT AND MATERIALS

In most nuclear transfer studies, the microscope used is inverted with Numarski or Hoffman lenses. As shown in *Figure 3.1A*, a Nikon TE300 inverted microscope with Eppendorf micromanipulators is employed in our study although we have obtained similar results with a Narishige micromanipulation system. The Piezo-drill system PMM150 is made by Prime Tech. The microinjector is CellTram Vario from Eppendorf.

Culture medium and manipulation solutions

Preculture medium

The culture medium we used for preculture of oocytes and reconstructed embryos is CZB medium (21) with glucose (CZBG). A stock is

Table 3.3   Composition of the CZB stock

Chemicals	Weight (mg)/1000 ml
MiliQ water	990 ml
NaCl (S-5886)	4760
KCl (P-5405)	360
$MgSO_4 \cdot 7H_2O$ (M-1880)	290
EDTA.2Na (E-6635)	40
Na-Lactate (L-7900)	5.3 ml
D-glucose (G-6152)	1000
KH_2PO_4 (P-5655)	160

Note: The stock should be 1000 ml after the chemicals are dissolved in 990 ml of MiliQ water.

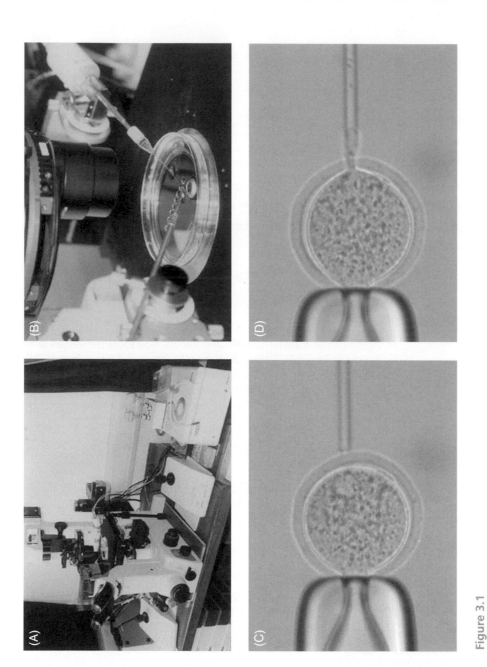

Figure 3.1

(A) The preparation of the piezo-drill microinjection system with NIKON TE300 inverted microscope and Eppendorf micromanipulator. (B) High magnification of the injection setting. (C) The MII spindle area of a B6D2F1 oocyte was viewed by NIKON inverted microscope with Normaski lenses. (D) The MII spindle was enucleated by an enucleation pipette with inner diameter of 8–10 μm.

Table 3.4   Composition of the CZBG

Chemicals	Weight (mg)/100 ml
CZB stock	99 ml
NaHCO3 (S-5761)	211
CaCl$_2$·2H$_2$O 100 × stock (C-7902)	1 ml
Pyruvate (P-4562)	3
Glutamine (GIBCO, 21051-024)	15
BSA (A-3311)	500

Table 3.5   Composition of the HCZBG

Chemicals	Weight (mg)/100 ml
CZB stock	99 ml
Hepes (H-4034)	476
NaHCO$_3$ (S-5761)	42
CaCl$_2$·2H$_2$O 100 × stock (C-7902)	1 ml
Pyruvate (P-4562)	3
Glutamine (GIBCO, 21051-024)	15
PVA (P-8136)	10

Note: Once all the chemicals have been added the pH of the medium should be adjusted to pH 7.4.

made which contains NaCl, KCl, NaH$_2$PO$_4$, Glucose, EDTA, MgSO4·7H$_2$O and lactic acid; 100 × CaCl$_2$·2H$_2$O is made at the same time. The composition of the stock is listed in *Table 3.3*. To make 100 ml of CZBG medium, NaHCO$_3$ is added followed by 1 ml of CaCl$_2$·2H$_2$O stock, then pyruvate, L-glutamine and BSA are included (*Table 3.4*).

In vitro manipulation medium

In vitro manipulation medium is Hepes buffered CZBG (HCZBG) medium. To make HCZBG medium, NaHCO$_3$ is reduced to 5 mM and 20 mM of HEPES is included, BSA is replaced by PVA with concentration of 0.1 mg/ml (*Table 3.5*).

Activation medium

The activation medium is calcium-free CZB medium containing 5 μg/ml cytochalasin B (CB) (Sigma) and 10 mM strontium (Sigma). To make activation medium, 900 μl of calcium-free CZBG medium (*Table 3.6*) is mixed with 100 μl of 100 mM strontium stock and 5 μg of cytochalasin B in 10 μl of DMSO. The activation medium made in this way initiates proper activation of reconstructed oocytes and no precipitation occurs during activation treatment. One percent DMSO in the activation medium has been demonstrated to be beneficial for cloned embryo development (22).

Suspension medium

The medium we used to suspend ES cells is 10% PVP in HCZB medium. To make a 10% PVP solution, 10 ml of HCZB medium is transferred to a Falcon 1007 petri-dish and 1 *g* of PVP is carefully poured onto the surface of HCZB. The dish is covered, sealed with parafilm and stored in the refrigerator for at least 2 days before filtering through a 0.8 μm syringe filter. This procedure allows the PVP to be dissolved totally and will not damage the cell or oocyte during micromanipulation.

Table 3.6   Composition of the calcium-free CZBG

Chemicals	Weight (mg)/100 ml
CZB stock	100 ml
NaHCO$_3$ (S-5761)	211
Pyruvate (P-4562)	3
Glutamine (GIBCO, 21051-024)	15
BSA (A-3311)	500

Methods

Protocol 3.1: Preparation of donor cells

ES cells were cultured in Glasgow MEM (GMEM) media (GIBCO) supplemented with 15% heat-inactivated FCS, 1000 units of leukemia inhibitory factor/ml (ESGRO) and the following reagents: 0.1 mM nonessential amino acids (GIBCO), 0.1 mM β-mercaptoethanol, 1 mM sodium pyruvate (GIBCO), 2 mM L-glutamine (GIBCO). ES cells were cultured to 80–90% confluence and 16 hours before experiments FCS concentration was reduced to 5%. The ES cells cultured in this way have shown better results in embryo development compared to cells cultured to low confluence in our study (11). The hybrid ES cells used in our study are established from *in vivo* blastocysts of C57BL/6J crossed with 129/SV.

NOTES

Protocol 3.2: Oocyte collection

In most nuclear transfer studies, B6D2F1 (C57BL/6J × DBA/2) female mice have been used as donor oocytes. Eight- to ten-week-old female mice were injected with 5 IU of pregnant mare serum gonadotrophin (PMSG) (Calbiochem) and 5 IU of hCG (Sigma) 44–48 h after PMSG injection. The oviducts of the female mice were collected 13–15 h after hCG injection. Cumulus–oocyte complexes were released into HCZB medium containing 300 IU/ml hylaronidase (ICN). After removal of cumulus cells, the cumulus-free oocytes were precultured in CZB medium before enucleation.

NOTES

Protocol 3.3: Enucleation

One drop of 10% PVP and several drops of HCZB medium containing 5 µg/ml of CB are aligned in a Falcon 1029 cover dish covered with mineral oil (Fisher). A group of 15–20 oocytes (depends on enucleation speed) are transferred to a drop of HCZB-CB media. For enucleation a pipette with inner diameter of approximately 10 µm is used. The pipette used for piezo-drill microinjection is made differently to the sharpened conventional injection pipette, and is cut flat. As shown in *Figure 3.1*, the MII spindle area can be detected as a translucent area under Numarski optics. After several piezo pulses, the enucleation pipette penetrates the zona and the spindle area is touched by the pipette. Then the spindle is sucked by the pipette and removed smoothly (*Figure 3.1*). Ten percent PVP is used to wash the enucleation pipette to inhibit sticking. Normally, 20 oocytes can be enucleated by an experienced person in less than 10 minutes. The enucleated oocytes are washed and kept in CZB culture drops before injection.

NOTES

Protocol 3.4: Microinjection of ES cell nuclei into enucleated MII oocytes

The confluent ES cells cultured in 5% FCS are trypsinized and washed in GMEM culture media. After centrifugation, the cell pellet is suspended in a small amount of 5% FCS media. To make the injection dish, several drops (15–20 μl/drop) of 10% PVP and HCZB are placed onto a cover slip. One drop of cell suspension is mixed thoroughly with a drop of 10% PVP media. An injection pipette with inner diameter of approximately 5 μm (depends on cell type) is used to pick up a donor ES cell nuclei. A small cell with diameter less than 10 μm is sucked into and out of the pipette to separate the nucleus from the cytoplasm. A piezo pulse may be employed to assist the pipette to break the cell membrane. After 4–6 ES cell nuclei are lined up inside the pipette, this should be moved to the injection drop containing a group of 10–15 oocytes. The nuclei are injected one by one into the enucleated oocytes. The setting of frequency and intensity of the piezo pulse to break the oolemma should be as low as possible, to minimize damage to the oocyte or cell nucleus. In order to provide a stable hydraulic column the pipette is backfilled with mercury. The quantity of mercury backfilled in the pipette and the size of the pipette both affect the piezo setting. Less mercury will produce much stronger power and make it easier for the pipette to break the membrane. A smaller size of pipette will break the membrane more easily. A small size of pipette filled with a small amount of mercury is therefore ideal for microinjection of ES cell nuclei. To improve the survival rate after injection, the pipette should push the membrane inwards while applying gentle suction. Then employ the piezo pulse to break the membrane. Once the membrane is broken, inject the cell nucleus into the cytoplasm with a small quantity of PVP. The injection pipette is then withdrawn quickly (which will help to improve the survival rate).

NOTES

Protocol 3.5: Activation of reconstructed oocytes

After incubation in HCZB media for several minutes (to allow membrane recovery), the injected oocytes are transferred back to CZB medium and cultured for 1–3 h before activation treatment. During this period of time, chromosome condensation and microtubule organization as a spindle should be observed (11). This phenomenon might be critical for further development of cloned embryos. After 1–3 h of culture, the oocytes are transferred to calcium-free CZB medium containing 5 µg/ml of Cytochalasin B and 10 mM of strontium for 5–6 h. Cytochalasin B inhibits pseudo polar body release and makes the oocyte preserve a diploid genome. Strontium has been reported to mimic sperm to activate mouse oocytes. Pseudo-pronuclei formation will be observed after activation treatment.

NOTES

Protocol 3.6: Cloned embryo culture and transfer into surrogate mothers

The activated oocytes with pseudo-pronuclei formation are collected and cultured in CZB medium or M16 (or KSOM or Witten's Medium). At the 4-cell stage, the embryos are transferred to a fresh culture drop till morulae/blastocyst stage prior for embryo transfer. Normally, after 72–76 h of culture from onset of activation, the embryos develop to morulae and early blastocyst stage. Five to ten embryos are transferred to each uterine horn of 2.5 days post coitum (dpc) pseudo-pregnant mice. The recipient mothers are killed at 19.5 dpc and the pups quickly removed from the uteri. After cleaning fluid away from their air passages, the pups are kept in a warm box supplied with oxygen. Surviving pups are raised by fostering onto lactating mothers.

NOTES

Protocol 3.7: Micromanipulation technique and preparation of tools

Micromanipulation tools are critical to successful enucleation and injection of cell nuclei into oocytes. After cutting on the microforge, the pipette should be checked under the inverted microscope to ensure that the end of the pipette is flat. The size of enucleation pipette is not critical and inner diameter ranges from 8–12 μm. Pipettes should be bent to a 20–40° angle by the microforge and then backfilled with mercury to the desired volume (1–2 mm long). The capacity of the pipette to make a hole in the zona depends on the piezo pulse setting and on the quantity of mercury backfilled in the pipette. The piezo pulse should be set as low as possible and a frequency setting of 3–5, intensity of 2–3, is normally sufficient to allow the pipette to penetrate the zona easily. The size of injection pipette depends on the cell type used for injection and an inner diameter of 5 μm is ideal to pick up G1 stage ES cell nuclei. As with enucleation pipettes this instrument is cut flat on the microforge. The piezo pulse setting frequency 1 and intensity 1 is sufficient to allow the pipette to break the oocyte membrane and inject the cell nucleus into the oocyte. To inhibit stickiness of the pipette during injection, ES cells are suspended in 10% PVP medium and some PVP will be injected into the oocyte with the cell nuclei. The amount of PVP injected affects the oocyte's survival rate and should be minimized. PVP is toxic to oocytes and better development of reconstructed embryos has been shown when a lower percentage of PVP is used to suspend donor cells (18).

NOTES

Protocol 3.8: Alternative method for murine nuclear transfer

The protocols above describe piezo-drill assisted nuclear injection. While this is not the only way to generate reconstructed embryos, it is well suited to small cells that may be difficult to fuse by other methods. Recently, mice have been successfully produced by electrofusion of tail tip cells (fibroblast cells) into enucleated oocyes (23). M-phase ES cells have also been fused with enucleated oocytes by sendai virus (24,25) and cloned pups have been produced. The fusion method is normally used for relatively large cells because if piezo microinjection is employed the pipette is bigger and fewer oocytes will survive after injection. It is much easier to use larger cells than small cells for fusion. As an alternative method to piezo-drill micromanipulation, the fusion method is relatively easily mastered.

NOTES

Gene targeting in sheep

4

A. John Clark and Wei Cui

4.1   Introduction

Pronuclear injection remains the mainstay for transgenic livestock production (1) and is now routine in a variety of species. However, there has been little improvement in the efficiency of the technique over the years, with only 2–3% of injected eggs giving rise to transgenic offspring. In addition, the multicopy nature of most transgenic loci, coupled with the random nature of the site of integration, may give rise to unpredictable levels of expression (2). Most importantly, direct pronuclear injection provides no realistic means to modify endogenous genes in a targeted manner.

In contrast, gene targeting in the mouse using embryonic stem (ES) cells has been enormously successful. By exploiting the capacity of these cells to undergo homologous recombination with exogenous DNA and contribute to the germ line, the introduction of precise, targeted changes into the mouse germ line is now routine (reviewed by (3)). The importance of this approach was understood at an early stage by those working with livestock, but ES cells capable of contributing to the germ line have not been isolated (4). This precluded the development of cell-based transgenesis and, consequently, gene targeting in livestock. The ability to clone animals by nuclear transfer (NT) from cultured somatic cells (5,6), however, gave rise to an alternative route to germ line modification applicable to many species (7). In this approach gene targeting is carried out in somatic cell lines prior to NT. Targeting vectors are designed to modify the gene of interest by homologous recombination. Enrichment of the targeting event is achieved by antibiotic selection and the targeted cell lines are then used in NT procedures (8,9).

Until recently there was little information on the frequency of homologous recombination in primary somatic cells and this was mainly restricted to human cells (10). Because primary somatic cells have a limited lifespan in culture, genetic modification and the preparation of the cells for NT must be accomplished before they senesce or enter crisis and transform. Lifespan in culture is determined by several parameters, including the culture conditions (11,12), age of the donor animal (13), and tissue and species of origin (14).

This chapter describes the use of primary cell cultures from sheep fetuses and their use in gene-targeting experiments and NT. The growth characteristics of these cell populations in terms of the requirements for introducing and selecting targeted events, and expanding targeted cell populations are described. Promoterless targeting vectors were constructed to disrupt the

Gene Targeting and Embryonic Stem Cells, Alison Thomson and Jim McWhir
© 2004 Garland Science/BIOS Scientific Publishers.

ovine *GGTA1* (α1,3-galactosyltranserase) genes and *PrP* (prion protein) gene. Using a high-throughput screening protocol, targeting events were detected at these loci but the limited proliferative capacity of the primary cells in culture was a significant impediment to overall success. A strategy to overcome this limitation by expressing the telomerase reverse transcriptase (*TERT*) gene to bypass senescence is described.

4.2 Growth characteristics of primary somatic cells

Successful nuclear transfer from cultured cells has only been achieved in livestock using primary somatic cells, a fundamental characteristic of which is their limited lifespan. We have estimated that achieving targeted genetic modifications requires at least 45 population doublings to proceed from isolation of primary cells from fetal tissue, through the targeting process to preparing targeted cells for nuclear transfer (NT) (7). Therefore, we set out to characterize the growth properties of fetal cells derived from different breeds of sheep, and determine whether their proliferative capacity could be improved by modifying the culture conditions (15).

Fetuses were isolated from Black Welsh, Finn Dorset and Shetland breeds of sheep that had been naturally mated. Following disaggregation of the carcass, individual cells and clumps of cells attached to the gelatin-coated tissue culture plastic and most cultures grew to confluency after 3–7 days. Most cells from each culture were cryopreserved for use in future gene-targeting experiments. The remaining cells were used to determine the maximum proliferative capacity by serial passage on standard tissue culture plastic, in an atmosphere composed of 75% N_2, 20% O_2 and 5% CO_2. Based on morphological appearance, fibroblastic cells predominated after about the third passage ($\sim$12 doublings) (*Figure 4.1A*). A surprising observation was the large variation seen both between cultures derived from the same breed and between different breeds. For example, BW6F2 divided $\sim$120 times at intervals of $\sim$20 hours but BWF1 senesced after less than 50 populations doublings (PD), each taking $\sim$40 hours to complete (*Figure 4.1B*). We have now investigated the growth characteristics of many primary sheep cultures and have shown that proliferative capacity can vary over a range of 30–120 population doublings.

Several studies have suggested that growth of cells in an atmosphere with reduced oxygen content may increase the rate and extent of proliferation (11,12). We compared the characteristics of three sheep fetal lines grown in 5% or 20% O_2 but failed to detect any growth advantage of bulk cultures in the low-oxygen atmosphere (*Figure 4.1C*).

4.3 Construction and transfection of gene-targeting vectors

To construct targeting vectors we used *GGTA1* or *PrP* DNA probes to screen genomic libraries prepared from Black Welsh sheep cells (line BWF1) to isolate the relevant genomic clones (*Figure 4.2*).

It has become increasing clear that efficient gene targeting in somatic cells requires powerful strategies to select for the targeted cells (10). The

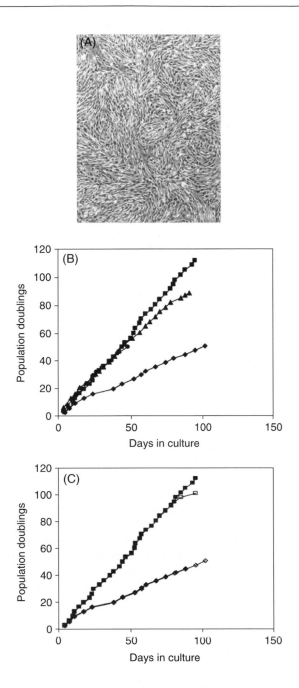

Figure 4.1

Growth characteristics of fetal fibroblasts in culture. Fibroblastic morphology of a confluent monolayer of cells at passage 4–6 cultured from a day 35 sheep fetus (**A**). Cells were isolated from 35-day-old carcasses and passaged continually in an atmosphere of 75% N_2, 20% O_2 and 5% CO_2 (**B**), or 90% N_2, 5% O_2 and 5% CO_2 (**C**) to senescence. Sheep cultures from Black Welsh (BW6F2,■ BWF1,▲), Finn Dorset (7G65F4,♦). Taken from Denning *et al.* (15) with permission.

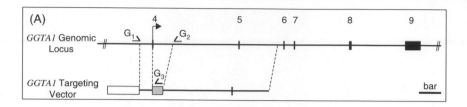

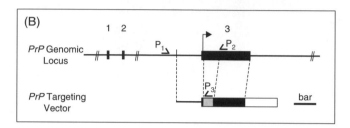

Figure 4.2

Organization of the genomic loci of (**A**) ovine *GGTA1*, (**B**) ovine *PrP* and the promoterless targeting vectors used for disruption. Numbering of the exons in *GGTA1* is based on the mouse; translation initiates in exon 4 and terminates in exon 9. The coding sequence of *PrP* is entirely within exon 3. G_1–G_3 and P_1–P_3, location of PCR primers. Black boxes represent exons, hatched boxes represent *neo-pA* sequence, and the open box represents pBlueScript sequence. Scale bar represents 2 kb. Taken from Denning *et al.* (9, 15) with permission. © Nature Publishing Group.

promoterless method has been reported to increase the ratio of targeted to nontargeted (random integration) clones by 100–500 fold because the selectable marker will only be expressed if integration occurs in frame and downstream of a transcriptionally active gene (10). It does, however, require that the target genes are expressed in the cell line being targeted. Using Northern analysis, both *GGTA1* and *PrP* transcripts were readily detectable (*Figure 4.3A*). In these replacement vectors the neomycin phosphotransferase (*neo*) gene was placed directly adjacent to the initiation codon of the targeted genes and a section of the downstream sequence was deleted. In correctly targeted clones, this strategy places the *neo* gene under the control of the endogenous promoter of either the *GGTA1* or *PrP* genes (*Figure 4.2*).

4.4 Detecting targeted cells

The *GGTA1* and *PrP* targeting constructs were transfected into early-passage fibroblasts. The targeting efficiency was anticipated to be low, so a high-throughput strategy to detect correctly targeted colonies was used. Following transfection, cells were seeded into a 96-well plate format and G418 selection was applied. Drug-resistant colonies were derived at an efficiency of ~1 in 10^{-4} to 10^{-5} cells. These were grown to subconfluence, then replica plated for cryopreservation and DNA analysis. Two independent PCR reactions to detect targeting events for each construct were used (*Figure 4.2*). One reaction

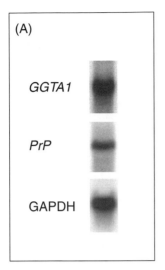

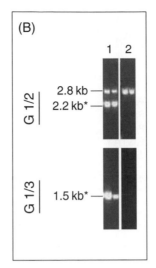

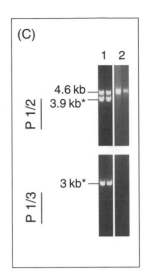

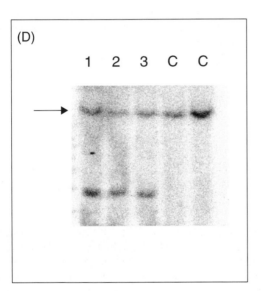

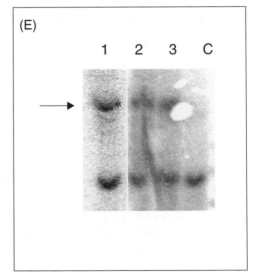

Figure 4.3

Northern blotting *of GGTA1* and *PrP* genes expressed in sheep fetal fibroblasts (**A**). Ovine coding sequence probes were used for *GGTA1* and PrP and murine coding sequence probe was used for the GAPDH control. DNA was isolated from *neo^R* colonies of transfected sheep and analyzed by PCR (see *Figure 4.2* for location of primers). (**B**) Ovine *GGTA1* PCR (lanes in duplicate). Panel 1, targeted cells, panel 2, labeled nontargeted parental cells. (**C**) Ovine *PrP* PCR (lanes in duplicate). The asterisks indicate the amplified fragments indicative of correct targeting. After nuclear transfer DNA was isolated from cloned fetuses and lambs and probed by Southern blotting with probes for *GGTA1* (**D**) or PrP (**E**). Lanes 1–3, targeted clones and C, controls. The arrow shows the correctly targeted alleles. Taken from Denning *et al.* (9). © Nature Publishing Group.

Table 4.1 Targeting efficiencies in ovine fetal fibroblasts

Breed	Locus	Drug resistant colonies	Detected targeting events[a]	Targeting frequency[b]	Absolute targeting frequency[c]
Sheep					
Black Welsh	PrP	856	55 (50)	6.4%	27.5×10^{-7}
Finn Dorset	PrP	917	14 (12)	1.5%	23×10^{-7}
Shetland	PrP	1241	19 (ND)	1.5%	6.3×10^{-7}
Black Welsh	GGTA1	1467	11 (3)	0.8%	2.8×10^{-7}
Finn Dorset	GGTA1	1485	45 (25)	3.0%	11.3×10^{-7}

The total number of detected targeting events is indicated[a] and the proportion of these that contained targeted and nontargeted cells (mixed colonies) is shown in parentheses. ND, not done. Targeting frequency[b] describes the ratio between detected targeting events and drug resistant colonies. Absolute targeting frequency[c] is targeting efficiency per cell basis. Taken from Denning *et al*. (15) with permission.

amplified from genomic DNA external to the left arm of homology into *neo* coding sequence and indicated whether targeted cells were present. In contrast, the second reaction was designed to detect both alleles, wild-type and targeted, resulting in fragments of different sizes (*Figure 4.2*). Moreover, since targeted and nontargeted alleles should be present in equimolar ratios, so producing bands of similar intensity, it was possible to determine with a reasonable degree of confidence whether a colony contained only targeted cells or was contaminated by additional nontargeted cells. The determination of the purity of colonies by Southern blotting was not attempted because it was not possible to isolate sufficient quantities of DNA from individual colonies routinely.

Using this PCR approach, targeting events at both the *GGTA1* (*Figure 4.3B*) and at the PrP (*Figure 4.3C*) locus were detected. The overall targeting frequency ranged from 0.8% to 9% (targeted to nontargeted cell ratio) and corresponded to an absolute targeting frequency of 2.8×10^{-7} to 27.5×10^{-7} on a per cell basis (*Table 4.1*). There was a large variation both between the targeting efficiencies in the different cell lines (*Table 4.1*) and between experimental groups using the same line (data not shown).

Although many of the sheep colonies were targeted, a high proportion were mixed; i.e. they contained targeted and nontargeted cells. Therefore, the data shown here represent the upper estimates of targeting efficiency. Nevertheless, a substantial number of cell clones that contained only targeted cells were identified. However, the majority of these pure clones senesced before they could be expanded sufficiently for NT. Hence the effective targeting frequency, in terms of generating useful targeted cells, was 10–30-fold lower than our estimate of actual targeting frequency; i.e. the effective frequency was in the order of 1 in 10^7.

4.5 NT with targeted sheep cells

Although most of the targeted colonies senesced, some of them did survive to be used in NT experiments (9), from which both fetuses (*GGTA1*

Table 4.2 Nuclear transfer from gene targeted primary cells

Stage of NT	Cells used for NT				
	3C6	5E1	4H2	7G65F	YH6
Embryos transferred into temp. recipients* (*in vitro* cultured)	87 (25)	0 (30)	92 (31)	55 (71)	273 (181)
Embryos recovered from temp. recipients	85	–	62	55	214
Morula or blastocyst[a]; *in vivo* (*in vitro*)	18 (7)	0 (3)	19 (8)	12 (27)	44 (3)
Embryos transferred to final recipients	18	3	23	33	43
Final recipients	12	3	17	18	28
Fetuses at day 35	7	2	4	8	18
Fetuses at day 60	5	0	2	5	8
Lambs at birth; live (dead)	0	0	0 (1)	0 (2)	3 (1)
Lambs alive at 1 week	0	0	0	0	1

Data are shown for various cultures: 3C6 and 5E1 (*GGTA1* correctly targeted), 4H2 (randomly integrated *GGTA1* targeting vector) and 7G65F (untransfected cells) were of Finn Dorset origin; YH6 (*PrP* correctly targeted) was of Black Welsh origin. Poll Dorset oocytes were used as recipient cytoplasts throughout. [a]Reconstructed embryos were transferred to temporary recipients, unless number of oocytes recovered was low or fusion could not be seen and *in vitro* culture (additional embryos shown in parentheses) was adopted. Taken from Denning *et al.* (9). © Nature Publishing Group.

and *PrP*) and live-born lambs (*PrP*) were produced (9). These animals were generated using standard NT protocols developed at Roslin. This involves serum starving the cells in culture for 3–5 days to induce quiescence prior to transfer into enucleated MII oocytes (5,6). The oocyte and cell are then joined by electrofusion of their cell membranes and the couplets transferred into temporary recipients. Morulae and blastocysts are recovered from the temporary recipient and transferred into final recipients. A high incidence of pre- and perinatal losses was encountered during these experiments and no animals survived beyond 2 weeks (*Table 4.2*). In the only other report to describe gene targeting by nuclear transfer in sheep (8), the birth and survival of two sheep carrying a specific gene insertion was reported. In this study there was a high incidence of pre- and perinatal loss compared to experiments using unmodified early-passage fetal fibroblasts (5). It seems highly likely that the overall efficiency of NT is reduced further by the stringent selection and extended culture required to isolate gene-targeted cells.

4.6 The control of replicative senescence

The parameters of cell growth and targeting efficiency on primary cells described above show that it is just about feasible to generate targeted sheep using a combination of gene targeting and NT. One of the key factors limiting success is, clearly, the replicative lifespan (16) of the donor cells used in these experiments. This has a dramatic effect on the targeting frequencies and is almost certain to impact on the efficiency of NT, notwithstanding the fact that successful cloning from senescing cells has been accomplished in cattle (13).

The replicative lifespan of cells can be extended in a number of ways. For example, tumor cell lines are immortal but these have lost the normal controls over cell division and very often have chromosomal abnormalities. Transfection with viral genes such as T antigen or cellular genes such a *myc* or *ras* can be used to immortalize cells but, again, this results in the loss of normal growth control, genomic instability; cells transformed in this manner would certainly not be competent in nuclear transfer experiments. Attempts to clone from a transformed cell line in cattle have not proved successful (17). Recent discoveries on the basic mechanisms controlling lifespan in primary somatic cells have suggested an alternative route to bypass senescence. It is known that progressive loss of the telomeric repeats at the ends of chromosomes play an important role in determining the proliferative lifespan of somatic cells in culture (18). The telomere hypothesis of cellular senescence proposes that when the telomeres shorten to a critical limit, they lose their function, leading to chromosomal fusion and the activation of DNA damage pathways, leading to cell arrest or apoptosis that results in cell senescence.

Telomeres are elongated by a multisubunit enzyme complex, telomerase, comprising a reverse transcriptase catalytic component (TERT) inactive in normal somatic cells, but active in the germ line, certain stem cells and many tumor cells. Indeed this is why certain stem cells, such as mouse ES cells, are able to divide indefinitely.

4.7 Telomerase extension of proliferative lifespan

Sheep primary fibroblasts behave in culture very much like primary human cells and, for example, exhibit senescent specific markers at the end of their lifespan (*Figure 4.4A*). Like human cells, their proliferation is associated with the progressive shortening of the telomeres (*Figure 4.4B*). Sheep telomeres are somewhat longer than their human counterparts at early passage (20 kb vs 12 kb) but they shorten at a comparable rate.

Transfecting human primary cells with the hTERT gene reconstitutes telomerase activity, stabilizes the telomeres and extends their proliferative lifespan and, in effect, this overrides the 'mitotic clock' (19). The human TERT gene was introduced into primary sheep fibroblasts to see if it could similarly restore telomerase activity and maintain telomere length. A number of stably transfected primary fibroblast lines were generated expressing hTERT (20). The telomerase catalytic activity was measured using the 'TRAP' assay which provides an artificial primer sequence as a substrate for the telomerase complex to elongate. A number of TRAP-positive lines were identified showing that the human telomerase was capable of interacting with the sheep components of the telomerase complex, in particular the sheep template RNA (*Figure 4.5A*). The TRAP-positive cell lines were grown in culture and, eventually, it became clear that these lines had been immortalized. They grew for more than 450 population doublings (the experiment was stopped at this stage!) whereas the telomerase negative control cell lines senesced after no more than 60 doublings or so (*Figure 4.5B*). The

(A)

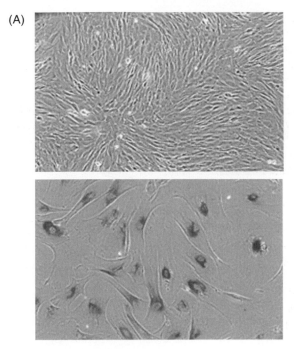

(B)

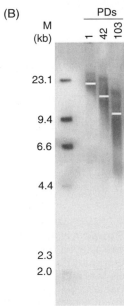

Figure 4.4

Proliferation and senescence of primary sheep fibroblasts. **(A)** Late passage fibroblasts contain a substantial proportion of senescing cells that can be visualized by a specific histochemical stain (lower panel): no senescing cells are present at early passages (upper panel). **(B)** Telomere shortening. Sheep telomeres are about 20 kb in length in early passage cells but shorten to an average of about 10 kb in presenescent cells. Taken from Cui *et al.* (20). © The American Society for Biochemistry and Molecular Biology.

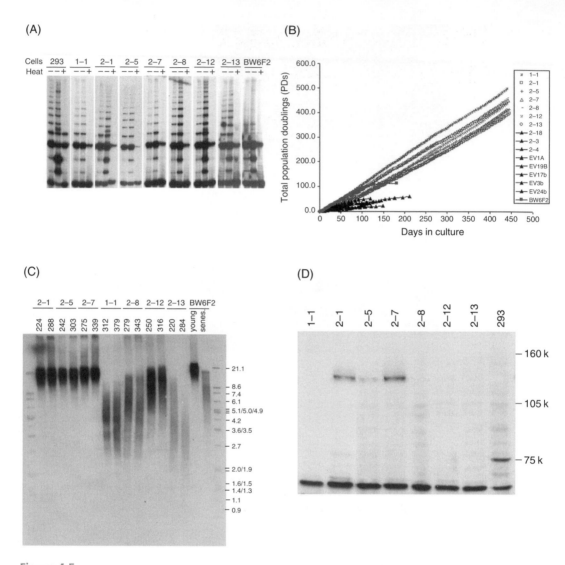

Figure 4.5

Reconstituting telomerase activity in sheep fibroblasts. (**A**) TRAP assay. A number of sheep primary cell lines were transfected with the hTERT gene. Cell extracts were assayed for telomerase activity by the telomere repeat amplification protocol (TRAP) assay. The primary parental cell line (BW6F2) does not express telomerase activity whereas the transfected cell lines and positive control, 293 (a human tumor cell line) all do, as evidenced by the ladder of bands generated by extending the telomere primer sequences in the TRAP assay. (**B**) Immortalization of sheep fibroblasts. TRAP-positive cell lines are immortal and continue to divide for more than 400 PDs. All the control cell lines senesced much earlier. (**C**) Telomere repeat fragment assays. This is a Southern blot using the telomeric repeat sequences as a probe. The TRF assays were carried out at various population doublings during the growth curves. Three of the lines maintained full-length telomeres which were the same size as those in the early passage parental cells (BW6F2) – even after 300 or more population doublings. Some of the other lines did exhibit telomere shortening but these stabilized at a set length characteristic of each line. (**D**) Western blot analysis. Only three cell lines exhibited high levels of TERT gene expression, as evidenced by Western blotting – these were the same three lines that fully maintained their telomeres. Taken from Cui *et al*. (20). © The American Society for Biochemistry and Molecular Biology.

telomere lengths were measured in these cell lines. A number of them showed no telomere shortening whatsoever; even after 450 doublings the telomeres were the same length as they were in early passage cells (*Figure 4.5C*). Furthermore, these cells showed few if any chromosomal abnormalities after this very extended period in culture. Some of the other lines, however, did show telomere shortening and, eventually, some genomic instability. This was directly correlated with the level of hTERT gene expression (*Figure 4.5D*). Careful measurements of the hTERT mRNA levels in these cells showed that they required no more than 1–2 copies of hTERT mRNA per cell to fully stabilize the telomeres and prevent genomic instability (20).

Although it was clear that expression of sufficient levels of hTERT could immortalize these cells and maintain their telomeres, a key question was whether or not they had been transformed and had lost growth control. We investigated a number of classical parameters of transformation, including loss of contact inhibition and the acquisition of serum-independent growth. The lines all showed normal growth control after exceeding their natural lifespan, suggesting that the expression of telomerase was bypassing senescence without transforming them.

Finally, gene-targeting experiments were carried out in telomerase immortalized sheep cells and this showed that they exhibited comparable absolute targeting efficiencies to wild-type primary fibroblasts. The overall recovery of targeted clones was, however, much higher since they do not now senesce (D. Zhao and C. Denning, unpublished observations).

4.8 Nuclear transfer with telomerase immortalized cells

One of the telomerase-expressing sheep lines was used as a nuclear donor for NT after nearly 100 population doublings after transfection. This line, maintained full-length telomeres and a normal karyotype throughout the extended period of cell culture. Nuclei derived from these cells did support early embryonic and fetal development after nuclear transfer (21). Thus there was no significant difference in efficiency of preimplantation development between embryos derived by NT from hTERT-expressing cells, (4.4% blastocysts) and the parental, nonexpressing cells (5.4% blastocysts). However, no fetuses from the hTERT-expressing cells survived beyond day 40 (*Table 4.3*).

Five fetuses derived by NT from hTERT-expressing cells were aborted after they were judged to have died by ultrasound scanning. Postmortem examination showed that they all had undergone early organ development and eyes, ears, limbs and tails were clearly observed in some of them. These data showed that, nuclei from telomerase immortalized sheep fibroblasts retain a substantial degree of developmental plasticity and can be reprogramed by the oocyte to direct embryonic and early fetal development. Although the hTERT-expressing cells are capable of early postimplantation development, none of these fetuses developed further. This could be due to a number of possibilities including the integration of the hTERT sequences disrupting a gene important for later fetal development or the effect of sustained telomerase expression on later fetal development. Alternatively, the accumulated

Table 4.3 Development of nuclear transferred embryos from telomerized sheep fibroblasts

Donor cells	Passage (PDs)[d]	Reconstructed KCC[e, f]	Fused (%)[f]	Blastocysts (%)[g, f] recipients	No. of final	Pregnant (%)[h] born		Lamb born
						21 days	60 days	
BW6F2[a]	−3 (15)	648	354 (54.6)	19 (5.4)	19	4 (21.0)	2 (10.5)	1 (5.3)
YH6[b]	2 (45)	684	454 (66.4)	45 (9.9)	28	20 (71.4)	10 (35.7)	5 (17.8)
2–5[c]	25 (114)	714	373 (52.2)	16 (4.3)	15	7 (46.7)	0	0

Untransfected sheep fibroblasts.[a] *PrP* targeted BW6F2.[b] hTERT-expressing BW6F2.[c] Passage number of transfection is defined as 0 and PDs of primary cells = 0.[d] Reconstructed karyoplast–cytoplast complexes.[e] 2–5 vs YH-6 or BW6F2, p > 0.05 (Chi-squared).[f] Rates were calculated based on the numbers of fused embryos and final recipients respectively.[g, h] Taken from Cui *et al.* (21) with permission.

mutational load in these cells after the extended period of cell culture may have blocked later development.

4.9 Concluding remarks

Nuclear transfer now enables gene targeting in livestock species such as sheep. Although feasible the process is very inefficient and the already low efficiencies associated with cloning are compounded by the difficulties of gene targeting in primary somatic cells with a limited lifespan. This perhaps explains why there have been only two reports of successful gene targeting in sheep to date (8,9). The approach has been extended to pigs. Although the development of NT in pigs (22) lagged somewhat behind that of sheep, there are now a number of reports of gene-targeted pigs (23,24). A major objective has been to delete the *GGTA1* gene to generate pig tissues and organs that may be more suitable for transplantation into humans. Perhaps most impressive has been the recent description of homozygous knockout pigs lacking both copies of the gene (25). In this work the first allele was deleted in primary fibroblasts using a promoterless *neo* gene in much the same way to that described above for sheep. Cell lines were then established from these cloned pigs and retargeted with the same vector. In this case the selection criterion was based on sorting cells physically which did not express the *gal* epitope on their surface. Interestingly, the *gal*⁻ cells selected had not been targeted at the second allele but, rather, this had been inactivated by random mutation. This in many ways is preferable because it means that eventually homozygous knockout pigs lacking any foreign DNA such as the neomycin gene can be generated by breeding.

Notwithstanding this success in pigs, methods to extend lifespan of the donor cells used in NT may have important implications for the genetic engineering of livestock. This is particularly so because of the long generation intervals of these species. For example, if multiple genetic changes are required it will not be feasible to introduce these singly into the germ line and then combine them by breeding. Rather, it will be necessary to build on and develop the types of approach described above, to enable some or all of these changes to be engineered into a single cell line prior to cloning. Although full-term development was not achieved with the first NT experiments using telomerase immortalized fibroblasts, the fact that nuclei from these cells can be reprogramed to undergo embryonic and early fetal development is encouraging and this is a promising way to engineer the cells used in nuclear transfer to make them more tractable for genetic modification.

Acknowledgments

I would like to thank all my colleagues at the Roslin Institute who have contributed to this work, most of which has been previously published (8,9,20,21). This work was funded by the BBSRC and the Geron Corporation.

References

1. Hammer RE, Pursel VG, Rexroad C, *et al.* (1985) Production of transgenic rabbits, sheep and pigs by microinjection. *Nature* **315**: 680–683.
2. Clark AJ, Bissinger P, Bullock DW, Damak S, Wallace R, Whitelaw CB, Yull F (1994) Chromosomal position effects and the modulation of transgene expression. *Reprod Fertil Dev* **6**: 589–598.
3. Hooper ML (1992) Embryonal stem cells: introducing planned changes into the germ line. In: Evans HJ (ed) *Modern Genetics Volume 1*. Harwood Academic Publishers, Switzerland.
4. Stice SL (1998) Opportunities and challenges in domestic animal embryonic stem cell research. In: Clark AJ (ed) *Animal Breeding: Technology for the 21st Century*. Harwood Aacademic Press, Switzerland, pp 64–71.
5. Campbell KHS, McWhir J, Ritchie WA, Wilmut I (1996) Sheep cloned by nuclear transfer from a cultured cell line. *Nature* **380**: 64–66.
6. Wilmut I, Schnieke AE, McWhir J, Kind AJ, Campbell KHS (1997) Viable offspring derived from fetal and adult mammalian cells. *Nature* **385**: 810–813.
7. Clark AJ, Burl S, Denning C, Dickinson P (2000) Gene targeting in livestock; a preview. *Trans Res* **9**: 263–275.
8. McCreath KJ, Howcroft J, Campbell KH, Colman A, Schnieke AE, Kind AJ (2000) Production of gene-targeted sheep by nuclear transfer from cultured somatic cells. *Nature* **405**: 1066–1069.
9. Denning C, Burl S, Ainsley A, *et al.* (2001) Deletion of the α(1,3)galactosyl transferase *(GGTA1)* and prion protein *(PrP)* genes in sheep. *Nature Biotech* **19**: 559–562.
10. Sedivy JM, Dutriaux A (1999) Gene targeting and somatic cell genetics – a rebirth or a coming of age? *Trends Genet* **15**: 88–90.
11. Falanga V, Kirsner RS (1993) Low oxygen stimulates proliferation of fibroblasts seeded as single cells. *J Cell Physiol* **154**: 506–510.
12. Saito H, Hammond AT, Moses RE (1995) The effect of low oxygen tension on the in vitro-replicative life span of human diploid fibroblast cells and their transformed derivatives. *Exp Cell Res* **217**: 272–279.
13. Kasinathan P, Knott JG, Moreira PN, Burnside AS, Joseph Jerry D, Robl JM (2001) Effect of fibroblast donor cell age and cell cycle on development of bovine nuclear transfer embryos in vitro. *Biol Reprod* **64**: 1487–1493.
14. Rubin H (1997) Cell aging in vivo and in vitro. *Mech Ageing Dev* **98**: 1–35.
15. Denning C, Dickinson P, Burl S, Wylie D, Fletcher J, Clark AJ (2001) Gene targeting in sheep and pig primary fetal fibroblasts. *Cloning Stem Cells* **3**: 205–215.
16. Hayflick L, Moorhead PS (1961) The limited in vitro lifespan of human diploid cell strains. *Exp Cell Res* **25:** 585–621.
17. Zakhartchenko V, Alberio R, Stojkovic M, *et al.* (1999) Adult cloning in cattle: potential of nuclei from a permanent cell line and from primary cultures. *Mol Reprod Dev* **54**: 264–272.
18. Harley CB, Futcher AB, Greider CW (1990) Telomeres shorten during ageing of human fibroblasts. *Nature* **345**: 458–460.
19. Bodnar AG, Ouellette M, Frolkis M, *et al.* (1998) Extension of life-span by introduction of telomerase into normal human cells. *Science* **279**: 349–352.
20. Cui W, Aslam S, Fletcher J, Wylie D, Clinton M, Clark AJ (2002) Stabilisation of telomere length and karyotypic stability are directly correlated with the level of hTERT gene expression in primary fibroblasts. *J Biol Chem* **277**: 38531–38539.
21. Cui W, Wylie D, Aslam S, Dinnyes A, King T, Wilmut I, Clark AJ (2003) Telomerase immortalised fibroblasts can be reprogrammed by nuclear transfer to undergo early development. *Biol Reprod* **69**(1): 15–21.

22. Polejaeva IA, Chen SH, Vaught TD, *et al.* (2000) Cloned pigs produced by nuclear transfer from adult somatic cells. *Nature* **407**: 86–90.
23. Dai Y, Vaught TD, Boone J, *et al.* (2002) Targeted disruption of the alpha1, 3-galactosyltransferase gene in cloned pigs. *Nat Biotechnol* **20**: 251–255.
24. Lai L, Kolber-Simonds D, Park KW, *et al.* (2002) Production of alpha-1,3-galactosyltransferase knockout pigs by nuclear transfer cloning. *Science* **295**: 1089–1092.
25. Phelps CJ, Koike C, Vaught TD, *et al.* (2003) Production of alpha 1,3 galactosyltransferase deficient-pigs. *Science* **299**: 411–414.

13. Chuang, C.F. and Meyerowitz, E.M. (2000) Specific and heritable genetic interference by double-stranded RNA in *Arabidopsis thaliana*. *Proc. Natl Acad. Sci. USA* **97**: 4985–4990.

14. Smith, N.A., Singh, S.P., Wang, M.B. *et al.* (2000) Total silencing by intron-spliced hairpin RNAs. *Nature* **407**: 319–320.

15. Levin, J.Z., de Framond, A.J., Tuttle, A. *et al.* (2000) Methods of double-stranded RNA-mediated gene inactivation in *Arabidopsis* and their use to define an essential gene in methionine biosynthesis. *Plant Mol. Biol.* **44**: 759–775.

Protocols

Contents

Protocol 4.1: Isolation of sheep fetal fibroblasts

MATERIALS

1. Day 35 sheep fetal carcass
2. Dissecting tools: sterile forceps and scissors
3. Culture flasks and dishes
4. 30 ml universals
5. Culture medium:

    90% G-MEM (Sigma, G5154)

    10% FCS (Globe Farm, Gilford, Surrey, UK)

    2 mM L-glutamine (200 mM, Invitrogen, 25030024)

    1 mM sodium pyruvate (100 mM, Invitrogen, 11360039)

    1 × MEAA (100×, Invitrogen, 11140035)

6. 0.1% Gelatin (Sigma, G1890)
7. PBS (Invitrogen, 14190094)
8. Gentamicin (Sigma, G1264), prepare 50 mg/ml in PBS as stock solution. Dilute to 100 µg/ml in PBS or 50 µg/ml in the culture medium as working solution.
9. Trypsin-EDTA solution (1×, Sigma, T3924)

METHOD

1. Warm culture medium/gentamicin and PBS/gentamicin at 37°C and add 20 ml 0.1% gelatin to a T175 culture flask to leave it at room temperature
2. Get day 35 sheep fetal carcass and store individual carcass in universals containing 10–15 ml PBS/gentamycin. Do the cell isolation as soon as possible for the best results
3. Prepare 3 × 60 mm dishes. Two of them contain 2 ml PBS/gentamicin and one contains 2 ml trypsin
4. Transfer the carcass from the universal to the dish containing PBS/gentamicin to remove the blood and wash again in the other dish with PBS/gentamicin, then into the dish with trypsin
5. Cut the carcass finely and place the dish in a 37°C incubator for 5 minutes

6. Check the dish under a microscope and there should be lots of cells in suspension

7. Take a 5 ml pipette and add 3 ml of culture medium/gentamicin to the dish. Pipette it up and down a few times and then transfer this to a 15 ml tube

8. Take a 10 ml pipette and rinse out the dish with 9 ml of culture medium/gentamicin. Transfer it to the tube as well. Pipette the tissue/cell suspension up and down in the tube

9. Allow the suspension to settle for 2–3 minutes. The larger pieces of tissues should settle to the bottom

10. Transfer the supernatant to a 50 ml tube, then add a further 6 ml medium to the original tube

11. Repeat steps 10–11 two to three times

12. Spin the tube at 1000 rpm for 5 minutes

13. Remove the gelatin from the flask and add 20 ml culture medium/gentamicin

14. Remove the supernatant from the tube and resuspend the cell pellet in 20 ml culture medium/gentamicin, then transfer the cell suspension to the flask (40 ml total volume). Grow cells at 37°C, 5% CO_2

15. After 24 hours, replace the medium with fresh culture medium without antibiotics

NOTES

Protocol 4.2: Electroporation and selection

MATERIALS

1. Plasmid DNA: suitable targeting plasmids with selectable markers (e.g. *neo* or *pac*)

2. QIEAXII Gel Extraction Kit (QIAGEN)

3. Hemacytometer

4. 96-well plates (Sterilin, 3860-096)

5. Geneticin (Invitrogene, 10131019), diluted to the appropriate concentration in the culture medium

6. 2 mg/ml puromycin, stock solution (Sigma, P8833), diluted to the appropriate concentration in the culture medium

7. Multichannel pipette and reagent reservoirs (Fisher)

8. Electroporation system (Bio-Rad, Gene Pulser II Apparatus)

METHOD

1. Preparation of plasmid DNA: Linearize the plasmid with an appropriate restriction enzyme. Purify the linearized fragment with phenol/chloroform or QIEAXII Gel Extraction Kit. 10 µg of purified DNA fragment is required for each electroporation

2. Sheep fetal fibroblasts were grown in culture medium to about 90% confluent in a T75 (or T175) flask

3. Trypsinize the cells with 2–5 ml trypsin

4. Count cell number. Transfer 5×10^6 cells to a 15 ml tube

5. Spin the cells at 1000 rpm. Aspirate the supernatant and resuspend the cells in 700 µl PBS

6. Prepare purified 10 µg DNA fragment to a volume of 100 µl with PBS

7. Transfer 700 µl of cells and 100 µl of DNA to an electroporation cuvette and electroporate the cells at 125–250 µF/250–400 V

8. After the electroporation, transfer the cells with a Pasteur pipette to sterile bottle containing 200 ml culture medium. Mix well

9. Aliquot 100 µl (2.5×10^3 cells) per well to 96-well plates

10. Grow at 37°C, 5% CO_2 for 24 hours before adding selection to the medium

11. At subconfluence, resistant colonies were trypsinized and replica-plated to two 96-well plates

12. One plate will be lysed for DNA analysis and the other will be cryopreserved

NOTES

Protocol 4.3: Colony picking

MATERIALS

1. Trypsin-EDTA solution (1×, Sigma, T3924)

2. 96-well plates (sterilin 3860-096)

3. PBS (Invitrogen, 14190094)

4. Culture medium

5. DMSO (Sigma)

METHOD

1. Circle the wells containing resistant colonies. Aspirate medium from the wells and wash cells with 70 μl PBS

2. Add 30 μl trypsin into each well and incubate at 37°C for 5 min

3. Add 80 μl of medium to the above well after trypsinization

4. Label two 96-well plates, one for culture and the other for freezing

5. Transfer 50 μl trypsinized cells into each well for each plate

6. Add 100 μl selection medium to each well in the culture plate and incubate in 37°C incubator with 5% CO_2. Feed the cells every 3–4 days

7. Add 100 μl of culture medium containing 10% DMSO to each well in the freezing plate and freeze it at −80°C after wrapping it with bubble wrap

NOTES

Protocol 4.4: Preparation of genomic DNA for PCR

MATERIALS

1. PBS (Invitrogen, 14190094)

2. Lysis buffer (50 mM Tris-HCl pH 8, 20 mM EDTA, 100 mM NaCl, 0.3% SDS, 10 mg/ml proteinase K)

3. 96-well plates (sterilin 3860-096)

4. 70% Ethanol

5. TE buffer (10 mM Tris-HCl, 1 mM EDTA, pH 8)

METHOD

1. Once cells grow to confluence, aspirate the medium and wash with PBS

2. Add 50 μl lysis buffer to each well

3. Incubate at 37°C o/n

4. Transfer the lysed cells from cell culture plate to a 96-well plate and add 50 μl isopropanol

5. Shake the plate for 10 min on a plate shaker and spin for 20 min at 3000 rpm

6. Flick to paper towels and tap gently

7. Add 50 μl 70% ethanol and spin for 5 min at 3000 rpm

8. Drain as before and air dry

9. Add 50 μl TE buffer and leave at 4°C o/n

10. Shake the plate for 30 min next day to get DNA suspension

11. 1 μl of DNA is required for each PCR reaction

NOTES

Protocol 4.5: Resuscitation of frozen cells from duplicate plate

METHOD

1. Warm culture medium to 37°C and add 2 ml to each well of a 24-well plate

2. Frozen plates are only removed from freezer right before resuscitation

3. Add 200 ml warm medium to each well and gently pipette up and down, then transfer to 24-well plate containing 2 ml medium

4. Grow cells in 37°C, 5% CO_2 incubator

NOTES

RNA interference

Florence Wianny, Roberta J. Weber and
Magdalena Zernicka-Goetz

5

5.1 Introduction

Gene targeting, in which DNA constructs are introduced into specific sites
in the genome of ES cells, and subsequently animals, has become a powerful tool for elucidating gene function by generating animals with loss of
function – so-called 'knockouts'. Until recently, homologous recombination
was the only method for gene targeting in mammals. But this method is fasti-
dious and costly and, with the exception of heterozygous 'knockouts', is not
readily adaptable to analysis of the effects of partial down-regulation of gene
expression. Conventional plasmid-based targeting requires the generation
of complex targeting constructs, the selection of homologous recombination events in ES cells, the production of chimeric mice after microinjection
of modified ES cells into blastocysts, and the establishment of pure breeding
lines, before obtaining homozygous mutant mice. An alternative tool for
ablating gene function in eukaryotes discovered in the past few years utilizes
double-stranded RNAs (dsRNAs). This new method, called 'RNA interference', or RNAi, could well compete in many instances with homologous
recombination to ablate gene function in mammalian cells.

RNAi was first discovered in *C. elegans*. In this species, antisense RNAs
were classically used to ablate gene function (1). In 1998, Andrew Fire and
Creg Mello made the unexpected observation that contaminating dsRNA
present in the single-stranded antisense RNA preparations was at least ten
fold more potent at ablating gene function than were either sense or antisense RNAs (2). Furthermore, exposure of an adult worm to only a few
molecules of dsRNA per cell triggered gene silencing throughout the animal
and also in its progeny. This system has now been adapted to large-scale
proteomic analysis in *C. elegans* (3–6).

Since this discovery, a flood of papers has appeared which describe the
application of this technology in many other systems. It has become clear
that dsRNA-induced gene silencing as a phenomenon is present in evolu-
tionary diverse organisms such as plants, nematodes, fungi, trypanosomes,
Drosophila melanogaster, and vertebrates (7–16).

5.2 The mechanism

Although the picture for the mechanism of RNAi remains incomplete,
convergent observations in diverse organisms suggest that a conserved

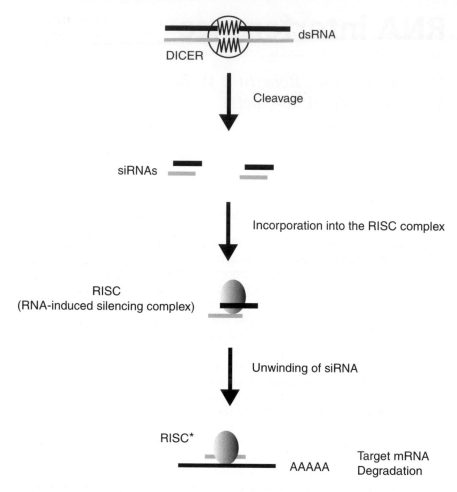

Figure 5.1

The model for RNA interference: dsRNA are processed by the DICER enzyme into ~22-nucleotide siRNAs, which are incorporated into a multicomponent nuclease, the RISC complex (17). RISC is activated from a latent form, containing a double-stranded siRNA to an active form, RISC*, after unwinding of siRNAs (21). RISC* uses the sequence of the single-stranded siRNA as a guide to identify and destroy homologous mRNAs (19,22).

mechanism is at the core of the different dsRNA gene-silencing phenomena (described in *Figure 5.1*). This begins with the conversion of dsRNA into small RNAs (approximately 22–25 nt) termed small interfering RNAs (siRNAs), by an RNAse III family nuclease, DICER (17). This conversion requires additional cofactors that may recruit the dsRNA to DICER or stabilize the siRNA product (13). These native siRNAs are incorporated into a protein complex called RNA-induced silencing complex (RISC). The active RISC complex uses the antisense strand of the siRNAs for targeting and destroying the cognate mRNAs (18–22). Cleavage seems to be endonucleolytic, and occurs only in the region homologous to the siRNA. Activated

RISC complexes can regulate gene expression at many levels: they certainly act by promoting RNA degradation and translational inhibition, and they probably also target chromatin remodeling.

5.3   RNAi in mammalian somatic cells

In invertebrates and plants, dsRNAs of variable lengths (generally of about 500–700 bp) were shown to be potent triggers of RNAi. And it was clear that RNAi would prove a powerful tool for manipulating gene expression in these species. However, there were some impediments to the use of this approach in mammals. Indeed, it was long known that the presence of extremely low concentrations of dsRNA in viral infections triggers the interferon response, part of which is the activation of a dsRNA-responsive protein kinase (PKR) (23,24). This enzyme phosphorylates and activates translation factor EIF2α in response to dsRNA. In addition, dsRNA activates the 2′5′ oligoadenylate synthetase, which causes mRNA degradation by RNase L. The consequence is a general blockage of protein synthesis, which in turn triggers cell death via apoptosis. Nevertheless, some investigators have found that a few mammalian cell lines could evade the toxic effect induced by long dsRNA (25,26). Also, the existence of a strong conservation of some genes involved in the RNAi pathway, like *Dicer* or *Argonaute* gene family members, also predicted that some steps of RNAi were active in mammalian cells (17,27).

5.3.1   Small interfering RNA (siRNAs)

A decisive clue to applying RNAi in mammalian cells came from the breakthrough discovery that dsRNAs of 21–23 nucleotides length, which are generated from cleavage of long dsRNA, were at the heart of RNAi machinery (28). Such small dsRNAs designed to mimic the product of the DICER enzyme, induced efficient sequence-specific mRNA degradation in lysates prepared from Drosophila embryo extracts (18,29). This result led to the idea that one way to avoid the nonspecific dsRNA responses observed in mammalian cells was to create dsRNA <30 bp in length. And indeed, when chemically synthesized siRNAs were introduced into mammalian cells, they appeared to evade the interferon response and induced sequence-specific gene silencing (30,31). Within a few months of this discovery, numerous reports appeared demonstrating that siRNA induced sequence-specific gene silencing in many different mammalian cell types (see 16 for review). When introduced into cells, the siRNAs assemble with proteins of RISC, which then guides target mRNA cleavage.

5.3.2   Chosing the best siRNA

The rules about how to design the most effective siRNA are to date very imprecisely defined. A general guideline has been published by Tuschl (the sirna user guide: http://www.mpibpc.gwdg.de/abteilungen/100/105/sirna.html).

Commonly used siRNAs resemble the DICER processing products of long dsRNAs that normally initiate RNAi: they are 21–23 bp long, with characteristic 2- to 3-nucleotide 3′overhangs on each strand, and 5′-phosphorylated termini. Blunt-ended siRNA duplexes or siRNAs lacking a 5′ phosphate group do not efficiently trigger gene silencing *in vitro* and *in vivo* (21,30,32,33). The GC content of the sequences should be between 30% and 70%. It is not known if any mRNA region is the most optimal for siRNA targeting. Nevertheless, the sequence of the siRNA should be carefully chosen, in order to avoid complementarity, and thus interference with an undesired mRNA. One or two bp mismatches between siRNA and mRNA dramatically reduce gene targeting and silencing (34). However, genes with high homology, or members of the same family could be affected by the same dsRNA. In general, it seems preferable to avoid regions around the translation start site as well as 3′ and 5′ UTR regions, which might bind RNA regulatory proteins. It is better to test several siRNA for targeting the same mRNA, as several siRNAs targeting different regions of the RNA can give different results (35).

5.3.3 Production of siRNA *in vitro*

Chemically synthesized siRNAs are widely used for gene targeting in mammalian somatic cells. However, one disadvantage of these siRNAs is that they can be expensive when purchased as 'ready-to-use' RNAs. Several groups have thus devised strategies to synthesize short RNAs *in vitro* (*Figure 5.2*). Three different strategies have been used so far:

1. The direct production of siRNAs, where siRNA sense and antisense strands are transcribed by individual promoters. These two strands are then annealed to produce a duplex of siRNA (36,37).
2. The production of short hairpin RNAs (shRNAs), where siRNAs are expressed as fold back stem loop structures (38).
3. The production of siRNAs after nucleolytic processing of long dsRNAs by a purified bacterial RNaseIII enzyme (39,40). This last approach may be the most effective way to produce siRNA *in vitro*, since it gives rise to several different siRNAs targeting the same mRNA. However, this strategy will probably necessitate a purification step to remove contaminating long dsRNA, and a careful check of interference specificity, as these complex siRNA populations could contain molecules interfering with other nontargeted mRNAs.

5.3.4 Towards a long-lasting effect

There are two limitations of siRNA gene-targeting experiments in mammalian cells. The first limitation is the transient nature of siRNA. Indeed, in mammalian cells, RNAi does not seem to involve an amplification process to confer RNAi potency and longevity as is seen in organisms such as worms and plants. The amount of protein from the targeted gene in siRNA-treated cells recovers between 5 and 7 days after siRNA transfection (7–10 rounds of cell divisions), and after 4 days when injected in the adult

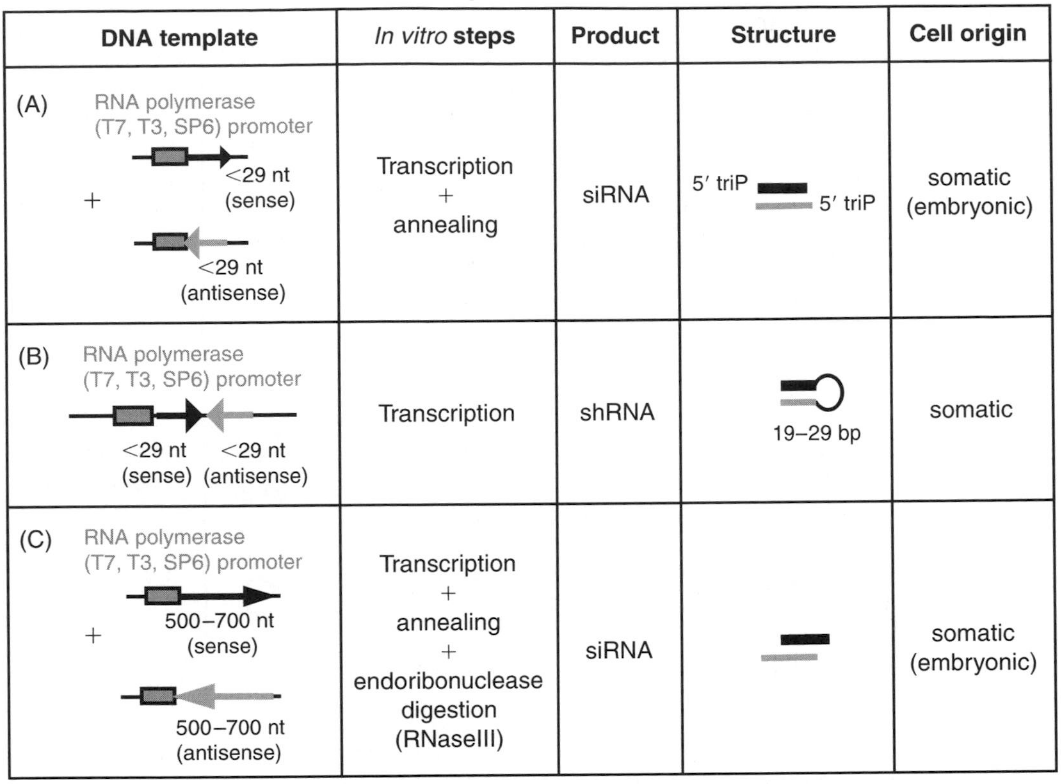

	DNA template	*In vitro* steps	Product	Structure	Cell origin
(A)	RNA polymerase (T7, T3, SP6) promoter + <29 nt (sense) / <29 nt (antisense)	Transcription + annealing	siRNA	5′ triP / 5′ triP	somatic (embryonic)
(B)	RNA polymerase (T7, T3, SP6) promoter <29 nt (sense) <29 nt (antisense)	Transcription	shRNA	19–29 bp	somatic
(C)	RNA polymerase (T7, T3, SP6) promoter + 500–700 nt (sense) / 500–700 nt (antisense)	Transcription + annealing + endoribonuclease digestion (RNaseIII)	siRNA		somatic (embryonic)

Figure 5.2

Different approaches to produce siRNA *in vitro*: (**A**) Sense and antisense siRNAs can be transcribed from separate promoters and annealed *in vitro*. (**B**) RNA polymerases T7 or T3 can be used to direct the transcription of small inverted repeats separated by a spacer region of varying lengths. The resulting RNAs form hairpins containing <29 nt stems that match the target sequence. (**C**) siRNAs are produced after digestion of long dsRNAs (synthesized and annealed *in vitro*) with a purified *E. coli* RNase III.

mouse (41). It is possible to extend the RNAi effects by inducing cell growth arrest, supporting the view that the loss of effectiveness of RNAi is due to siRNA dilution during cell divisions rather than degradation of siRNAs. Thus, the effect will largely depend on the type of cell, on its proliferation rate, but also on the stability of the targeted mRNA and protein. The second limitation of the use of siRNAs is the requirement for chemical or enzymatic synthesis of siRNAs before application to the cell. These siRNA produced *in vitro* are indeed easily subject to degradation by RNases.

Because of these two important limitations, several groups have developed *in vivo* expression constructs for small dsRNA triggers in mammalian cells. To date, two major approaches have been used and are similar to those used for *in vitro* production of siRNA (*Figure 5.3*).

1. In the first approach, two RNA polymerase promoters are placed in tandem (42,43), or on two separate vectors (37) to direct transcription of

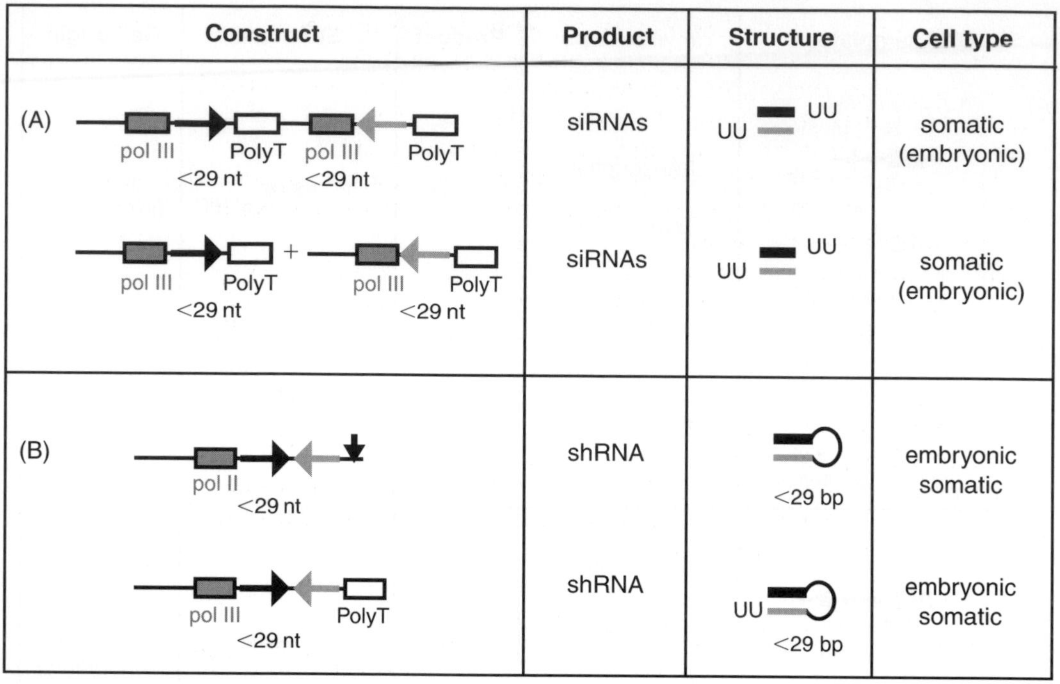

	Construct	Product	Structure	Cell type
(A)	pol III PolyT pol III PolyT <29 nt <29 nt	siRNAs	UU ▬ UU ▬	somatic (embryonic)
	pol III PolyT + pol III PolyT <29 nt <29 nt	siRNAs	UU ▬ UU ▬	somatic (embryonic)
(B)	pol II <29 nt	shRNA	▬◯ <29 bp	embryonic somatic
	pol III PolyT <29 nt	shRNA	UU ▬◯ <29 bp	embryonic somatic

Figure 5.3

Producing siRNA *in vivo*: **(A)** Sense and antisense strands can be transcribed *in vivo* from two independent RNA polymerase III promoters, which are either placed in tandem on the same plasmid, or on two separate plasmids. **(B)** shRNA can be produced *in vivo*, after transcription of a sequence containing the sense and antisense strands placed in tandem downstream of the RNA polymerase pol II. In this case, the plasmid is linearized before transfection into the cells. The sequence can also be placed downstream of a RNA polymerase pol III promoter. In this case, the transcription terminates when the RNA polymerase recognizes a sequence of four thymidines.

sense and antisense strands of a small RNA. These small RNAs are believed to form a duplex *in vivo*, similar to chemically synthesized siRNA.

2. In the second approach, two promoters are used to direct expression of small inverted repeats (19–29 nt) to create short hairpin RNAs, with 19–29 bp stem and 3–9 nt loop. These hairpins may then be processed by DICER to active siRNAs *in vivo* (44).

In these approaches, RNA polymerase III (pol III) promoters (U6 and H1) are mostly used because they have advantages over other promoters: almost all their elements are located upstream of the transcribed region, so that almost any inserted sequence shorter than 400 nucleotides can be transcribed. They are therefore suited for the expression of 21 long nucleotide, or short RNA stem loops. Moreover, RNA transcription terminates at a defined site, when Pol III encounters a run of four or five thymidines. The use of pol III promoters is thus recommended. However, its use is not compulsory: RNA polymerase II-promoter has also been used successfully for triggering siRNA production *in vivo* (45–47).

Even when using the approaches described above, suppression of the protein is still transient. However they have advantages over siRNAs synthesized *in vitro* as they can be combined with strategies for stable expression (34). The construct can also be placed under the control of inducible promoters to interfere with specific gene expression in a spatio-temporal manner *in vitro* and *in vivo* (48,49).

Applying one of these approaches for long-term effects can be time consuming. This drawback has to be seriously considered, especially when taking into account the rapidity of siRNA-directed mRNA degradation, which is often observable in less than 20 h (50). Most of the time, the effects of siRNA will persist for sufficient time to observe the loss of function phenotype, and long-lasting siRNA production is not necessary. Whether or not it is necessary will depend on the final aim. If one would like to knock out a gene, for example only in the brain tissue, this would need the incorporation of the hairpin construct downstream of an endogenous promoter whose activity is restricted to the brain.

5.3.5 Delivery and recipients

DsRNA (long dsRNA, siRNA and plasmid encoding for dsRNAs) can be transfected using classical methods of transfection. It is advisable to test which transfection reagent is best suited for a given cell line. Lipophilic agents, like Oligofectamine™, lipofectine and lipofectamine that carry the dsRNA across the membranes, are mostly used because they often appear less toxic than other methods of transfection (51,43). Electroporation can also be used to introduce siRNA into mammalian cells, and in embryos (39,45,50,52). This method usually gives high siRNA transfection efficiency (>90%), but generally appears cytotoxic (more than 50% of the cells can die). Viral vectors have also been used for introducing siRNA in cells that are otherwise difficult to transfect. For example, retroviruses and adenovirus have been shown to be effective vehicles for delivering siRNAs into cells (46,53). Lentiviral vectors should also be perfect vehicles for the introduction of DNA templates expressing siRNA *in vivo*, and especially suited to nonproliferating cells.

Recent reports show that siRNA or small hairpin RNAs can also be delivered *in vivo*, directly in the organs of an adult mouse (liver or heart), or injected by high pressure into the mouse tail vein (41,54). The effects observed after injections in the mouse tail vein are specific in a variety of organs, such as the liver, spleen, lung, kidney and pancreas (41).

RNAi is applicable for use in many different types of mammalian cells. Hcla cells, which are well known for their ease of transfection, are commonly used for studying the effects of gene ablation by siRNA. Other cell lines of different origins (mouse, monkey, human) have also been successfully used: mouse embryonic fibroblasts, 3T3, COS, Caco2, 293. DsRNAs have even been introduced in cell types in which genetic manipulations are especially difficult like mammalian postmitotic neuronal cells (30,55). The efficiency of transfection will depend not only on the cell line, but also on the passage number and the confluency of the cells.

5.3.6 Control and evaluation of gene inhibition

Once the dsRNA corresponding to the gene of choice has been introduced into the cells, the specificity of gene ablation should be carefully checked. The introduction of unrelated dsRNAs should neither give rise to any decrease in the level of the targeted gene product, nor should it induce any toxic effects. Moreover, the effects of the introduction of the targeting dsRNA on general gene expression have to be monitored.

For monitoring gene ablation, cells should be examined for levels of targeted mRNA, (by quantitative RT-PCR, or Northern blot), and protein (by Western blot or immunocytochemistry). Down-regulation of gene products can also be assessed by examining the phenotype.

It is important to take into account the stability of the targeted protein (as stable proteins require a longer period of exposure to siRNAs in order to be knocked down than less stable ones), the regulation of the protein, and also its concentration in the cell (highly expressed transcripts might be more difficult to knock down than rare ones). All these parameters will alter the outcome of the experiment.

It is advisable to carefully check the results, as substantial variability in gene ablation can be observed, even within the same experiment. More-over, it seems that several siRNA duplexes have to be tested for targeting the same gene. Indeed, shifting the siRNA target site by only a few nucleotides may result in different degrees of gene silencing (35). Thus, if knockdown of a gene is not observed, it is often advisable to test another siRNA sequence that targets a different region of the gene.

5.4 RNAi in mammalian embryonic cells

5.4.1 Mouse oocytes and embryos

Homologous recombination, the classical method used for inhibiting gene expression in the mouse embryo is very powerful – but not in all circum-stances. When the aim is to disrupt the function of the gene of interest in specific cells of the early embryo and/or at defined times, this technology generally cannot be used, as it eliminates gene function equally throughout the embryo. We say 'generally' because a type of homologous recombina-tion involving the Cre/lox system of deletion after the recombination event can be used under tissue- and temporal-specific promoters, but is extremely laborious and requires multiple lines of Knockout and Cre mice (56). Moreover, none of these methods can address the function of a maternally expressed gene during oocyte maturation or early embryonic development, when the ablation of this gene is deleterious in homozygous mutants. A few years ago, RNAi appeared to be the 'saving' tool, which was potentially capa-ble of overcoming all these limitations. Indeed, soon after Fire and colleagues described this new technology in worms, we found that mouse oocytes and preimplantation embryos were unexpectedly competent for RNAi (57). Moreover, they clearly lacked a vehement response to the introduction of long dsRNA, which was in contrast to what had generally been observed in somatic mammalian cells until then. The RNAi technology has since been improved and used by several other laboratories for inhibiting gene expression in the mouse oocyte and embryo, using either long dsRNAs or siRNAs (summarized in *Table 5.1*; see below for the application of the siRNA technology).

Introduction of dsRNA in the embryo

In *C. elegans*, RNAi can be provoked by injection of dsRNA into the gonad, but also through feeding either of dsRNA itself, or of bacteria engineered to express it (2,58). Mouse oocytes apparently lack this uptake mechanism, or if it is present, it is very inefficient. Culture of oocytes in medium containing dsRNA does not reduce the amount of the targeted mRNA (59).

Consequently in mammals the ways to introduce DNA or mRNA must be a little more 'sophisticated'. The techniques used so far are mentioned in *Table 5.1*. A classical way to introduce constructs into the egg is microinjection. We had developed in our laboratory approaches for microinjection of synthetic mRNAs in mouse oocytes and preimplantation embryos for directing gene expression (60). This technique was adapted to microinjection of dsRNAs, and described in Protocols 5.1 and 5.2.

However, microinjection demands micromanipulation skills, it is time consuming and requires costly apparatus. Furthermore, it is not efficient as a means of introducing dsRNA into all of the blastomeres after the two-cell stage, and only a single embryo can be microinjected at any one time.

Thus with the attempt to develop the use of RNAi in a more high-throughput manner, we have adapted the electroporation technique to the introduction of dsRNA into mouse oocytes and embryos (45). In order to increase the development potential of the embryos, and to decrease the cytotoxic effects of electroporation, the zona pellucida was not completely removed from the embryos, but only 'weakened' by brief treatment with Tyrode's solution. In these conditions, the efficiency of electroporation is variable but could often reach 100%. Furthermore, it did not affect the developmental potential of the embryos. This technique allows the simultaneous delivery of dsRNA into several embryos, and can be used for later stage embryos (i.e. four cell stage embryos). It now seems that it can also be successfully used for embryos at the postimplantation stages (39,52).

Specific silencing with long dsRNA

To determine whether the RNAi technology might be applicable in the mammalian embryo, we produced long dsRNA *in vitro* by synthesizing sense and antisense strands from individual promoters, and annealed equimolar quantities of these two strands *in vitro*. The purified preparations (500–700 bp long) were microinjected into the cytoplasm of immature oocytes, or one-cell stage embryos. As a proof of principle, we first targeted the marker gene *gfp*. We used a transgenic strain of mice that expresses a modified form of the green fluorescent protein (MmGFP) (60,61) from the ubiquitous elongation factor-1α (EF1α). This pilot series of experiments provided a rapid visual assay for evaluating the effect of dsRNA targeting *gfp*. We found that injection of *gfp dsRNA* in the one-cell stage embryo prevented the expression of *gfp* during preimplantation development. Similar

Table 5.1 Use of dsRNA in mouse oocytes and embryos

Cell type	Stage	dsRNA type	Length	Introduction	Target	Phenotype	Verif.	Reference
Oocyte	PRO I	In vitro synthesis and annealing	550 bp	M	c-mos	Parthenogenetic activation	WB	57
Oocyte	PRO I	dsRNA produced in vitro from PCR products	535 bp	M	c-mos	Inhibition of MAP kinase, Parthenogenetic activation	RT-PCR,	59
Oocyte	PRO I	dsRNA produced in vitro from PCR products	650 bp	M	Plat	No	RT-PCR, zymographic assay	59
Oocyte	PRO I	Hairpin dsRNA produced from inverted repeat	535 bp	M	c-mos	Inhibition of MAP kinase activity	MAP kinase activity	59
Oocyte	PRO I	In vitro synthesis and annealing	Full length ORF	M	MISS	Severe spindle defects	IF, comparison with morpholinos	75
Oocyte	PRO I	dsRNA produced in vitro from PCR products	550 bp	E	c-mos	Failure to arrest in metaphase II. Similar to c-mos-/-oocytes	nr	45
Oocyte	PRO I	SiRNA chemically synthesized and annealed in vitro	19 bp	M	c-mos	Failure to arrest in metaphase II. Similar to c-mos-/-oocytes	IF	74
Embryo Gfp	1 cell stage	In vitro synthesis and annealing	714 bp	M	gfp	Loss of green fluorescence	Confocal observation	57

System	Stage	Method	Size	Delivery	Target	Phenotype	Detection	Ref
Embryo	1 cell stage	In vitro synthesis and annealing	580 bp	M	E-cadherin	Severe pre-implantation defects. Similar to E-Cad-/-embryos	IF, WB	57
Embryo	1 cell stage	SiRNA chemically synthesized and annealed in vitro	19 bp	M	Oct4	Similar to oct4-/-embryos	IF	74
Embryo Gfp	1 cell stage	SiRNA produced in vivo from inverted repeat cloned downstream to the H1 promoter	21 bp	M	GFP	Loss of green fluorescence in all the organs and all developmental stages	WB	72
Embryo	1–2 cell stage	In vitro synthesis and annealing	Full length ORF	M	MISS	No	Co injected with MISS-GFP	75
Embryo Gfp	1–4 cell stage	dsRNA produced in vitro from PCR products	600 bp	E	gfp	Loss of green fluorescence	Confocal observation	45
Embryo	7.5 dpc	dsRNA produced in vitro from PCR products	500 bp	I+E	Otx2	Downregulation Reduction of forebrain and midbrain	ISH, IF, comparison with morpholinos	52
					Foxa2	Downregulation Loss of floorplate	nos	
Embryo	9.5 dpc	In vitro produced dsRNA digested with RNaseIII	15–40 bp	I+E (whole embryo)	LacZ gfp (from Tis21 locus)	βGal expression abolished in neuroepithelial cells. Decrease in green fluorescence (90%)	Co injected with βGal expression plasmid	39

MISS protein: 'MAPK-Interacting and Spindle-Stabilizing' protein.
Nd, not reported; PROI, prophase I; E, electroporation; I, injection; M, microinjection; IF, immunofluorescence; WB, Western blot; ISH, in situ hybridization.

results were obtained when endogenous genes were targeted in the early embryo. E-cadherin was severely knocked down after injection of the cognate dsRNA in the one-cell stage embryo. The treated embryos showed preimplantation defects, which were similar to those of null mutant embryos produced by homologous recombination. Similarly, we found that RNAi could also be used to unravel the function of a gene during oocyte maturation. This was validated using the same approach as above by utilizing another model gene, which had a characteristic knockout phenotype, *c-mos*. We and others had selected *c-mos* as a candidate gene because it is essential during oocyte maturation to maintain arrest at metaphase II (62). Oocytes lacking *c-mos* mature to metaphase II but then undergo spontaneous activation (63,64). And indeed we found that the injection of dsRNA directed towards *c-mos* faithfully phenocopied the *c-mos* null mutant, resulting in the specific reduction of the c-mos mRNA or protein. Consequently, this led to parthenogenetic activation of the oocytes (57,59).

These observations not only showed that RNAi was effective and specific in the mouse oocyte and embryo, but they also showed that the introduction of long dsRNAs did not produce any deleterious effects in the mammalian oocytes and embryos. Similarly, RNAi can be triggered in the mouse oocyte by production of long dsRNA *in vivo*, after microinjection of plasmids containing an inverted repeat of the gene of interest, driven by either the CMV promoter, or an oocyte-specific promoter 65).

5.4.2 RNAi in embryonic-derived cell lines with long dsRNAs

Several investigators attempted to extend previous findings in mouse embryos by searching for RNAi mechanisms in embryonic derived cell lines *in vitro*. They used two models:

1. Embryonic stem (ES) cell lines, derived from the preimplantation stage embryo.
2. Embryonic carcinoma (EC) cells derived from teratocarcinomas.

RNAi was first validated in ES and EC cells using long dsRNAs produced *in vitro*. It was found that these long dsRNAs triggered specific RNAi responses in ES and EC cells, without eliciting any deleterious effect on cell survival (49,66,67). This is in line with previous findings that these cells are deficient in some of the dsRNA or interferon activated enzymes and that induction of IFN genes by dsRNA or viral infection is impaired (68–71). Investigators mostly used dsRNAs that targeted transient expression of reporter genes (*gfp, luciferase, lacZ*). For example, cotransfection of EC cells with expression plasmids encoding *gfp* or *LacZ* together with the cognate dsRNA (500–700 bp in length), caused a specific decrease or elimination of the expression of the targeted gene (49,66). Similarly, transfection of ES cells with plasmids that direct the expression of firefly and *Renilla* luciferases in the presence of the cognate dsRNA specifically reduced the activity of the targeted enzyme (49).

dsRNAs have also been produced as long hairpin RNAs. These long hairpin RNAs were produced *in vitro*, using kits for large-scale RNA production. For example, Yang and colleagues cloned an inverted repeat of a long portion of the *gfp* gene (~550 nt long) under the control of the T7 RNA polymerase promoter. The long hairpin dsRNA was transcribed *in vitro*, and transfected into ES cells together with a plasmid encoding for GFP (67). The same authors also produced long hairpin dsRNA *in vivo*, after direct transfection of the same plasmid containing the inverted repeat, together with a plasmid encoding GFP and a plasmid encoding the T7 polymerase. These long hairpin dsRNAs, produced *in vitro* or *in vivo*, caused a specific decrease in GFP expression.

One report showed that long dsRNAs could also be used to target endogenous genes, like integrin α3 and β1 in EC cells (66).

However, RNAi was found to be more pronounced at early time points after transfections and decreased as ES cells divided, presumably due to the dilution of dsRNA (67). Stable silencing by long dsRNAs was recently achieved in ES cells by enforced expression of long hairpin dsRNA (49). ES cells were transfected with a plasmid containing a hairpin dsRNA targeting *gfp* and the *zeocine* resistance gene. Cells that had stably integrated the construct were selected by drug treatment and were amplified. These cells showed a stable and sequence-specific silencing of GFP expression.

However, one can speculate that the inhibition of gene expression by these long hairpin dsRNAs may become unspecific as ES cells differentiate. Indeed, the use of long dsRNAs for specific inhibition seems mainly restricted to cells with embryonic origins, i.e., cells of the preimplantation embryo, and ES/EC cells *in vitro*. RNAi is no longer potent when cells differentiate *in vivo*. We, and others, found that injection of *gfp* dsRNA in *gfp* transgenic zygotes caused an elimination of GFP expression until the blastocyst stage. However, the green fluorescence returned by the postimplantation stage (57,72). This is at the time when stem cells present in the ICM enter a differentiation program soon before gastrulation. During this time period, some 10–20 cells of the ICM also undergo a 100 fold increase in cell number. Thus, the disappearance of RNAi effects may also be the consequence of an important dilution of the injected dsRNA. Similarly, the effects of RNAi become unspecific when ES cell differentiation is induced *in vitro* (66,67).

The appearance of these unspecific effects placed a significant limitation on the utility of the long dsRNA approach in embryonic derived cells. Therefore, several groups developed the siRNA technology in mouse embryonic cells *in vitro* and *in vivo*.

5.4.3 siRNAs

siRNAs in ES cells

As expected from results obtained in somatic cells, chemically synthesized siRNAs were shown to be potent triggers of RNAi in ES cells. In our laboratory, we found that synthetic 21 bp siRNAs directed against *gfp* abolished the expression of GFP driven from an exogenous expression plasmid. Similarly, siRNAs caused a decrease in GFP fluorescence when introduced in ES cells that stably expressed GFP from an endogenous promoter (45). We also produced siRNA *in vivo*, using a plasmid encoding for short hairpin RNA corresponding to *gfp*, which was placed downstream of the CMV promoter. We found that transfection of this shRNA encoding plasmid led to an efficient and specific inhibition of GFP expression in ES cells. More importantly, both synthetic siRNAs and short hairpin RNAs were able to inhibit GFP expression in differentiated cells derived from ES cells. Thus, siRNAs enable similar interference effects in proliferative ES cells, as well as in their differentiated derivatives. However, the effect was still transient, and disappeared after few days.

In order to direct stable expression of siRNAs in ES cells, Carmell and colleagues used a transgenic approach to express a shRNA that targeted *Neil1*, a gene which may have a role in DNA repair. For this purpose, they cloned a small inverted repeat (19 nt) of the *Neil1* gene in a vector that carried the *neo* resistance gene and electroporated this vector in ES cells. ES cells that stably expressed the transgene were selected and several lines were amplified. The majority of these lines showed an 80% reduction in the level of Neil1 *mRNA* and protein, and an increase in their sensitivity to ionizing radiation. Moreover, these ES cells gave rise to differentiated derivatives that still expressed the shRNA (73).

siRNAs *in vivo*: from the embryo to the adult

The efficiency of siRNAs has first been validated in mouse oocytes and embryos by microinjecting chemically synthesized siRNAs. siRNAs directed against *c-mos*, in the oocyte, and against *Oct4* in the embryo, resulted in functional defects that were expected for the gene's inactivation (74,75). The siRNA technology was further developed by driving the expression of siRNAs in later stage embryos and in the adult mouse, using classical transgenesis approaches. Transgenes encoding shRNAs were microinjected in one-cell stage embryos and the effects of shRNAs were followed throughout development and in the adults. For example, Hasuwa and colleagues produced transgenic embryos expressing a shRNA directed against *gfp*. They found that this transgenically supplied shRNA could silence ubiquitously expressed *gfp* in the embryo, and in the adult mouse and rat (72). However, this standard transgenesis approach for stable gene silencing does not seem applicable in every instance. It was indeed found that shRNAs directed against a variety of other endogenously expressed genes did not give rise to any observable or reproducible phenotypes, although the presence of the transgenes driving shRNA expression could be detected in some animals (73). Germ line transmission of RNAi was validated using the ES cells that stably expressed shRNA directed against *Neil1* (73). These modified ES cells were microinjected into recipient blastocysts, which developed into fertile adult mice. It was found that the modified ES cells were able to contribute to the development of the host embryos, and to transmit the transgene in different organs of the chimeric mice, where the expression of the targeted gene was highly reduced. Numerous F1 progeny obtained after matings of these chimeric mice were shown to carry the shRNA-expressing vector. Furthermore, the expression of the targeted gene was greatly reduced in these F1 animals. This is the first demonstration that gene knockdown by RNAi can be transmitted to future generations.

This pioneering work opens the door to the development of strategies for inducible or tissue-specific ablation of genes *via* RNAi in the mouse. This prospect is especially appealing when embryonic lethality is the most likely outcome of gene disruption.

5.5 Conclusion

RNAi represents an exciting breakthrough in the quest to find an easy way to assess the function of genes in mammalian embryonic cells. This invaluable tool has numerous advantages compared to classical methods of gene knockout.

First, gene ablation can be targeted in specific cells of the early embryo and at defined times. This is an important point when considering the function of genes repeatedly used in space and time to direct developmental processes. In that case, inhibiting gene expression by homologous recombination, for example, will only allow the understanding of the first event.

Second, this technique allows the inhibition of maternally expressed products, which may mask the effect of the gene knockout. This provides a means of studying maternal effect of genes that show lethality in homozygous mutants. Third, in contrast to gene knockout by homologous recombination, RNAi can lead to moderate effects, referred to as gene 'knock down'. This is of particular interest, because it allows variable phenotypes to be obtained that are missed when the classical knockout technology is used. Hypomorph phenotypes can actually be the source of interesting information about the function of a gene. This could also bring some insights regarding the threshold level of gene function required to observe a mutant phenotype.

RNAi is also advantageous because groups of genes can potentially be rendered ineffective without the need for time-consuming crosses.

Finally, RNAi offers another big advantage in comparison to classical knockout technologies: it is a relatively cheap and quick method of gene ablation. The function of numerous factors can be screened very rapidly, and interesting ones can be selected for developing strategies for tissue-specific or inducible inhibition of gene expression in mice.

Although nothing is yet known regarding the effects of RNAi in other mammalian embryos, we can speculate that this powerful tool may be adapted to domestic species, in which it is difficult to create loss of function mutants.

References

1. Fire A, Albertson D, Harrison S, Moerman D (1991) Production of antisense RNA leads to effective and specific inhibition of gene expression in *C. elegans* muscle. *Development* **113**: 503–514.
2. Fire A, Xu S, Montgomery MK, Kostas SA, Driver SE, Mello CC (1998) Potent and specific genetic interference by double-stranded RNA in *Caenorhabditis elegans* [see comments]. *Nature* **391**: 806–811.
3. Fraser AG, Kamath RS, Zipperlen P, Martinez-Campos M, Sohrmann M, Ahringer J (2000) Functional genomic analysis of *C. elegans* chromosome I by systematic RNA interference. *Nature* **408**: 325–330.
4. Gonczy P, Echeverri C, Oegema K, *et al.* (2000) Functional genomic analysis of cell division in *C. elegans* using RNAi of genes on chromosome III. *Nature* **408**: 331–336.
5. Ashrafi K, Chang FY, Watts JL, Fraser AG, Kamath RS, Ahringer J, Ruvkun G (2003) Genome-wide RNAi analysis of *Caenorhabditis elegans* fat regulatory genes. *Nature* **421**: 268–272.
6. Kamath RS, Fraser AG, Dong Y, *et al.* (2003) Systematic functional analysis of the *Caenorhabditis elegans* genome using RNAi. *Nature* **421**: 231–237.
7. Misquitta L, Paterson BM (1999) Targeted disruption of gene function in Drosophila by RNA interference (RNA-i): a role for nautilus in embryonic somatic muscle formation. *Proc Natl Acad Sci USA* **96**: 1451–1456.
8. Bass BL (2000) Double-stranded RNA as a template for gene silencing. *Cell* **101**: 235–238.
9. Catalanotto C, Azzalin G, Macino G, Cogoni C (2000) Gene silencing in worms and fungi. *Nature* **404**: 245.
10. Cogoni C, Macino G (2000) Post-transcriptional gene silencing across kingdoms. *Curr Opin Genet Dev* **10**: 638–643.
11. Li YX, Farrell MJ, Liu R, Mohanty N, Kirby ML (2000) Double-stranded RNA injection produces null phenotypes in zebrafish [published erratum appears in *Dev Biol* 2000; **220**(2): 432]. *Dev Biol* **217**: 394–405.
12. Shi H, Djikeng A, Mark T, Wirtz E, Tschudi C, Ullu E (2000) Genetic interference in *Trypanosoma brucei* by heritable and inducible double-stranded RNA. *Rna* **6**: 1069–1076.
13. Hammond SM, Boettcher S, Caudy AA, Kobayashi R, Hannon GJ (2001) Argonaute2, a link between genetic and biochemical analyses of RNAi. *Science* **293**: 1146–1150.
14. Matzke MA, Matzke AJ, Pruss GJ, Vance VB (2001) RNA-based silencing strategies in plants. *Curr Opin Genet Dev* **11**: 221–227.
15. Wang Z, Englund PT (2001) RNA interference of a trypanosome topoisomerase II causes progressive loss of mitochondrial DNA. *EMBO J* **20**: 4674–4683.
16. McManus MT, Sharp PA (2002) Gene silencing in mammals by small interfering RNAs. *Nat Rev Genet* **3**: 737–747.
17. Bernstein E, Caudy AA, Hammond SM, Hannon GJ (2001) Role for a bidentate ribonuclease in the initiation step of RNA interference. *Nature* **409**: 363–366.
18. Tuschl T, Zamore PD, Lehmann R, Bartel DP, Sharp PA (1999) Targeted mRNA degradation by double-stranded RNA in vitro. *Genes Dev* **13**: 3191–3197.
19. Hammond SM, Bernstein E, Beach D, Hannon GJ (2000) An RNA-directed nuclease mediates post-transcriptional gene silencing in Drosophila cells. *Nature* **404**: 293–296.
20. Zamore PD, Tuschl T, Sharp PA, Bartel DP (2000) RNAi: double-stranded RNA directs the ATP-dependent cleavage of mRNA at 21 to 23 nucleotide intervals. *Cell* **101**: 25–33.
21. Nykanen A, Haley B, Zamore PD (2001) ATP requirements and small interfering RNA structure in the RNA interference pathway. *Cell* **107**: 309–321.

22. Martinez J, Patkaniowska A, Urlaub H, Luhrmann R, Tuschl T (2002) Single-stranded antisense siRNAs guide target RNA cleavage in RNAi. *Cell* **110**: 563–574.

23. Marcus PI (1983) Interferon induction by viruses: one molecule of dsRNA as the threshold for interferon induction. *Interferon* **5**: 115–180.

24. Clemens MJ (1997) PKR – a protein kinase regulated by double-stranded RNA. *Int J Biochem Cell Biol* **29**: 945–949.

25. Yi CE, Bekker JM, Miller G, Hill KL, Crosbie RH (2003) Specific and potent RNA interference in terminally differentiated myotubes. *J Biol Chem* **278**: 934–939.

26. Park WS, Miyano-Kurosaki N, Hayafune M, Nakajima E, Matsuzaki T, Shimada F, Takaku H (2002) Prevention of HIV-1 infection in human peripheral blood mononuclear cells by specific RNA interference. *Nucleic Acids Res* **30**: 4830–4835.

27. Nicholson RH, Nicholson AW (2002) Molecular characterization of a mouse cDNA encoding Dicer, a ribonuclease III ortholog involved in RNA interference. *Mamm Genome* **13**: 67–73.

28. Elbashir SM, Lendeckel W, Tuschl T (2001) RNA interference is mediated by 21- and 22-nucleotide RNAs. *Genes Dev* **15**: 188–200.

29. Elbashir SM, Martinez J, Patkaniowska A, Lendeckel W, Tuschl T (2001) Functional anatomy of siRNAs for mediating efficient RNAi in *Drosophila melanogaster* embryo lysate. *EMBO J* **20**: 6877–6888.

30. Caplen NJ, Parrish S, Imani F, Fire A, Morgan RA (2001) Specific inhibition of gene expression by small double-stranded RNAs in invertebrate and vertebrate systems. *Proc Natl Acad Sci USA* **98**: 9742–9747.

31. Elbashir SM, Harborth J, Lendeckel W, Yalcin A, Weber K, Tuschl T (2001) Duplexes of 21-nucleotide RNAs mediate RNA interference in cultured mammalian cells. *Nature* **411**: 494–498.

32. Chiu YL, Rana TM (2002) RNAi in human cells: basic structural and functional features of small interfering RNA. *Mol Cell* **10**: 549–561.

33. Schwarz DS, Hutvagner G, Haley B, Zamore PD (2002) Evidence that siRNAs function as guides, not primers, in the Drosophila and human RNAi pathways. *Mol Cell* **10**: 537–548.

34. Brummelkamp TR, Bernards R, Agami R (2002) A system for stable expression of short interfering RNAs in mammalian cells. *Science* **296**: 550–553.

35. Holen T, Amarzguioui M, Wiiger MT, Babaie E, Prydz H (2002) Positional effects of short interfering RNAs targeting the human coagulation trigger tissue factor. *Nucleic Acids Res* **30**: 1757–1766.

36. Donze O, Picard D (2002) RNA interference in mammalian cells using siRNAs synthesized with T7 RNA polymerase. *Nucleic Acids Res* **30**: e46.

37. Yu JY, DeRuiter SL, Turner DL (2002) RNA interference by expression of short-interfering RNAs and hairpin RNAs in mammalian cells. *Proc Natl Acad Sci USA* **99**: 6047–6052.

38. Leirdal M, Sioud M (2002) Gene silencing in mammalian cells by preformed small RNA duplexes. *Biochem Biophys Res Commun* **295**: 744–748.

39. Calegari F, Haubensak W, Yang D, Huttner WB, Buchholz F (2002) Tissue-specific RNA interference in postimplantation mouse embryos with endoribonuclease-prepared short interfering RNA. *Proc Natl Acad Sci USA* **99**: 14236–14240.

40. Yang D, Buchholz F, Huang Z, Goga A, Chen CY, Brodsky FM, Bishop JM (2002) Short RNA duplexes produced by hydrolysis with *Escherichia coli* RNase III mediate effective RNA interference in mammalian cells. *Proc Natl Acad Sci USA* **99**: 9942–9947.

41. Lewis DL, Hagstrom JE, Loomis AG, Wolff JA, Herweijer H (2002) Efficient delivery of siRNA for inhibition of gene expression in postnatal mice. *Nat Genet* **32**: 107–108.

42. Lee NS, Dohjima T, Bauer G, *et al.* (2002) Expression of small interfering RNAs targeted against HIV-1 rev transcripts in human cells. *Nat Biotechnol* **20**: 500–505.

43. Paul CP, Good PD, Winer I, Engelke DR (2002) Effective expression of small interfering RNA in human cells. *Nat Biotechnol* **20**: 505–508.
44. Paddison PJ, Caudy AA, Bernstein E, Hannon GJ, Conklin DS (2002) Short hairpin RNAs (shRNAs) induce sequence-specific silencing in mammalian cells. *Genes Dev* **16**: 948–958.
45. Grabarek JB, Plusa B, Glover DM, Zernicka-Goetz M (2002) Efficient delivery of dsRNA into zona-enclosed mouse oocytes and preimplantation embryos by electroporation. *Genesis* **32**: 269–276.
46. Xia H, Mao Q, Paulson HL, Davidson BL (2002) siRNA-mediated gene silencing in vitro and in vivo. *Nat Biotechnol* **20**: 1006–1010.
47. Zeng Y, Wagner EJ, Cullen BR (2002) Both natural and designed micro RNAs can inhibit the expression of cognate mRNAs when expressed in human cells. *Mol Cell* **9**: 1327–1333.
48. Miyagishi M, Taira K (2002) U6 promoter-driven siRNAs with four uridine 3' overhangs efficiently suppress targeted gene expression in mammalian cells. *Nat Biotechnol* **20**: 497–500.
49. Paddison PJ, Caudy AA, Hannon GJ (2002) Stable suppression of gene expression by RNAi in mammalian cells. *Proc Natl Acad Sci USA* **99**: 1443–1448.
50. McManus MT, Haines BB, Dillon CP, Whitehurst CE, van Parijs L, Chen J, Sharp PA (2002) Small interfering RNA-mediated gene silencing in T lymphocytes. *J Immunol* **169**: 5754–5760.
51. Chen Z, Indjeian VB, McManus M, Wang L, Dynlacht BD (2002) CP110, a cell cycle-dependent CDK substrate, regulates centrosome duplication in human cells. *Dev Cell* **3**: 339–350.
52. Mellitzer G, Hallonet M, Chen L, Ang SL (2002) Spatial and temporal 'knock down' of gene expression by electroporation of double-stranded RNA and morpholinos into early postimplantation mouse embryos. *Mech Dev* **118**: 57–63.
53. Brummelkamp TR, Bernards R, Agami R (2002) Stable suppression of tumorigenicity by virus-mediated RNA interference. *Cancer Cell* **2**: 243–247.
54. McCaffrey AP, Meuse L, Pham TT, Conklin DS, Hannon GJ, Kay MA (2002) RNA interference in adult mice. *Nature* **418**: 38–39.
55. Krichevsky AM, Kosik KS (2002) RNAi functions in cultured mammalian neurons. *Proc Natl Acad Sci USA* **99**: 11926–11929.
56. Lewandoski M, Martin GR (1997) Cre-mediated chromosome loss in mice. *Nat Genet* **17**: 223–235.
57. Wianny F, Zernicka-Goetz M (2000) Specific interference with gene function by double-stranded RNA in early mouse development. *Nat Cell Biol* **2**: 70–75.
58. Timmons L, Fire A (1998) Specific interference by ingested dsRNA. *Nature* **395**: 854.
59. Svoboda P, Stein P, Hayashi H, Schultz RM (2000) Selective reduction of dormant maternal mRNAs in mouse oocytes by RNA interference. *Development* **127**: 4147–4156.
60. Zernicka-Goetz M, Pines J, McLean Hunter S, Dixon JP, Siemering KR, Haseloff J, Evans MJ (1997) Following cell fate in the living mouse embryo. *Development* **124**: 1133–1137.
61. Zernicka-Goetz M, Pines J, Ryan K, Siemering KR, Haseloff J, Evans MJ, Gurdon JB (1996) An indelible lineage marker for Xenopus using a mutated green fluorescent protein. *Development* **122**: 3719–3724.
62. Gebauer F, Richter JD (1997) Synthesis and function of Mos: the control switch of vertebrate oocyte meiosis. *Bioessays* 19: 23–28.
63. Colledge WH, Carlton MB, Udy GB, Evans MJ (1994) Disruption of c-mos causes parthenogenetic development of unfertilized mouse eggs [see comments]. *Nature* **370**: 65–68.
64. Hashimoto N, Watanabe N, Furuta Y, *et al.* (1994) Parthenogenetic activation of oocytes in c-mos-deficient mice [see comments] [published erratum appears in *Nature* 1994; **370**(6488): 391]. *Nature* **370**: 68–71.

65. Svoboda P, Stein P, Schultz RM (2001) RNAi in mouse oocytes and preimplantation embryos: effectiveness of hairpin dsRNA. *Biochem Biophys Res Commun* **287**: 1099–1104.
66. Billy E, Brondani V, Zhang H, Muller U, Filipowicz W (2001) Specific interference with gene expression induced by long, double-stranded RNA in mouse embryonal teratocarcinoma cell lines. *Proc Natl Acad Sci USA* **98**: 14428–14433.
67. Yang S, Tutton S, Pierce E, Yoon K (2001) Specific double-stranded RNA interference in undifferentiated mouse embryonic stem cells. *Mol Cell Biol* **21**: 7807–7816.
68. Burke DC, Graham CF, Lehman JM (1978) Appearance of interferon inducibility and sensitivity during differentiation of murine teratocarcinoma cells in vitro. *Cell* **13**: 243–248.
69. Barlow DP, Randle BJ, Burke DC (1984) Interferon synthesis in the early post-implantation mouse embryo. *Differentiation* **27**: 229–235.
70. Francis MK, Lehman JM (1989) Control of beta-interferon expression in murine embryonal carcinoma F9 cells. *Mol Cell Biol* **9**: 3553–3556.
71. Harada H, Willison K, Sakakibara J, Miyamoto M, Fujita T, Taniguchi T (1990) Absence of the type I IFN system in EC cells: transcriptional activator (IRF-1) and repressor (IRF-2) genes are developmentally regulated. *Cell* **63**: 303–312.
72. Hasuwa H, Kaseda K, Einarsdottir T, Okabe M (2002) Small interfering RNA and gene silencing in transgenic mice and rats. *FEBS Lett* **532**: 227–230.
73. Carmell MA, Zhang L, Conklin DS, Hannon GJ, Rosenquist TA (2003) Germline transmission of RNAi in mice. *Nat Struct Biol* **21**: 21.
74. Kim M, Yuan X, Okumura S, Ishikawa F (2002) Successful inactivation of endogenous Oct-3/4 and c-mos genes in mouse preimplantation embryos and oocytes using short interfering RNAs. *Biochem Biophys Res Commun* **296**: 1372.
75. Lefebvre C, Terret ME, Djiane A, Rassinier P, Maro B, Verlhac MH (2002) Meiotic spindle stability depends on MAPK-interacting and spindle-stabilizing protein (MISS), a new MAPK substrate. *J Cell Biol* **157**: 603–613.
76. Yu JY, DeRuiter SL, Turner DL (2002) RNA interference by expression of short-interfering RNAs and hairpin RNAs in mammalian cells. *Proc Natl Acad Sci USA* **99**: 6047–6052.

Protocols

Contents

Protocol 5.1: Synthesis of dsRNA *in vitro*

TRANSCRIPTION OF SENSE AND ANTISENSE RNAS *IN VITRO*

General advice

RNAs are sensitive to cleavage by contaminating RNAses. Because RNAses are present on the skin, constant vigilance is required to prevent contamination of glassware and benchtops. Wear gloves, and use RNase-free glassware and plasticware. Set aside special electrophoresis devices for RNA loading. It is advisable for all chemicals to be reserved for work with RNA, and must be handled with disposable spatulas. Use as many disposable materials (tips, etc.) as possible.

DNA preparation

1. Cloning of the cDNA fragment into the expression vector. Insert the 500–700 bp cDNA in an expression plasmid downstream of the T7, T3 or sp6 polymerase promoters. For example, the plasmid pBluescriptSK can be used.

2. Linearize the plasmid (about 5–10 μg) with the appropriate restriction enzyme. Confirm that all the plasmid DNA has been converted from circular to linear molecules by analyzing an aliquot by gel electrophoresis.

3. Treat with proteinase K to remove all traces of restriction enzymes: Proteinase K ensures maximum purity of the vector DNA for efficient *in vitro* transcription. Transfer the linearized DNA to fresh Eppendorf tubes containing the following mixture:

Proteinase K (25 mg/ml)	2 μl
SDS (20%)	10 μl
DEPC-H$_2$O	to 360 μl

   Vortex to mix and and spin briefly. Incubate the reaction mixture for 20 min at 37°C.

4. To remove the proteinase K, extract the reaction mixture once with phenol/chloroform, and once with chloroform alone to remove all traces of phenol.

5. Transfer the aqueous phase to a fresh microfuge tube, and add 40 μl of 3 M Na acetate. Add 2.5 volumes of absolute ethanol.

6. Recover the DNA by centrifugation at maximum speed (minimum of 10 000 *g*) for 15 min in a microcentrifuge. Remove the supernatant and add 1 ml of 80% EtOH at room temperature to the tube. Centrifuge again for a further 2 min. Discard the supernantant, and dry the pellet.

7. Dissolve the pellet in autoclaved Ultra-pure/milliQ/18.2 MOhm water (not DEPC-treated, since traces of DEPC could prevent the activation of RNA polymerases).

8. Check the recovery of DNA by analyzing an aliquot by gel electrophoresis.

RNA transcription

Prepare single-stranded mRNAs with commercially available kits. We used Megascript kits, from Ambion. The kits are for large-scale *in vitro* synthesis of RNAs. The protocol is very straight-forward.

1. 1 µg of linearized DNA is mixed with ATP, UTP, CTP, GTP, 10× reaction buffer, and 10× enzyme mix plus the appropriate amount of RNAse free water.

2. Transcribe at 37°C for 2 h.

3. Digest away the DNA template by adding 2IU RNAse free DNAse I. Incubate for 20 min at 37°C. Add 115 µl DEPC-H_2O, and 15 µl 'stop solution' (5 M ammonium-acetate, 0.1 M EDTA). Extract the reaction with phenol/chloroform.

4. Purify the single-stranded RNA by adding 150 µl (1 volume) isopropyl alcohol. Precipitate by freezing at −80°C or overnight at −20°C. Resuspend the pellet with 10 µl H_2O.

5. Concentration of RNA can be checked by reading the UV absorbance at 260 nm, although these readings may not yield objective values, or by running 1 µl on a standard TAE agarose gel. Comparative analysis of the amount of RNA produced can be made by running it alongside 1/10th the amount of vector used (0.1 µg). The RNA intensity should be approximately 10× that of the DNA intensity.

If the RNA produced is to be used for injection *in vitro*, either as sense or antisense, then the RNA must be capped in order to ensure stability. One commercially available kit for transcribing and capping the RNAs is the Message Machine kit (Ambion).

All the RNAs are stored at −80°C.

DOUBLE-STRANDED RNA PREPARATION AND PURIFICATION

1. To anneal single-stranded sense and antisense RNAs, mix equimolar quantities of sense and antisense strands in the annealing buffer. For example, prepare the following mixture:

sense RNA (4 µg/µl)	4 µl
antisense RNA (4 µg/µl)	4 µl
2× annealing buffer	8 µl

Denature at 70°C for 10 min.

Incubate at 37°C for 1 h.

2× annealing buffer: 20 mM Tris pH 7.4, 0.2 mM EDTA

2. Remove the single-stranded RNAs by treatment with RNase T1 and RNase A:

Prepare the following RNase digestion mix:

Tris (1 M; pH 7.5)	35 μl
EDTA (0.5 M)	3.5 μl
NaCl (5 M)	21 mu;l
RNaseA*	0.35 μg
RNaseT1*	0.7 μg
mix sense +antisense RNAs	16 μl
DEPC-H₂O	to 350 μl

Preparation of RNase stocks:

Note: Use Ultra-pure/milliQ/18.2 MOhm water for all preparations.

RNase A: aliquot 1 ml of 10 mg/ml RNAse A into individual Eppendorf tubes. Boil these stock tubes for 1 h. Cool and dilute to 1 mg/ml, again in water. Store in 1 ml aliquots at –20°C.

RNase T1: dilute to 2 mg/ml in water, and stored at –20°C in 100 μl aliquots. RNaseA and RNase T1 can be thawed and frozen repeatedly without any loss of activity.

3. Vortex. Incubate for 30 min at 37°C.

4. Treat with proteinase K (see above)

5. Extract once with phenol/chloroform

6. Purify the dsRNA by precipitation with 40 μl Na-acetate (3 M) and 1 ml EtOH. Precipitate overnight at –20°C, or 15 min on dry ice. Spin at maximum speed (minimum of 10 000 **g**). Remove the supernatant and wash the pellet with 1 ml of 80% EtOH at room temperature. Dry the pellet.

7. Resuspend the pellet in 8 μl milliQ autoclaved water.

8. The proper annealing and concentration of dsRNA can be estimated by running 1 μl on a standard TAE agarose gel, in parallel to the corresponding single stranded sense and antisense RNAs, and by reading the UV absorbance at 260 nm.

ALIQUOTING AND STORING SSRNA AND DSRNAS IN 'MICROTUBES'

The volume to be microinjected into the oocytes and embryos are very small (~10 pl). Therefore, RNAs are aliquoted and stored into 'microtubes' which are made from 20 μl disposable glass capillaries (Supracaps) using the following method.

1. Cut the capillaries into 1 cm lengths with a diamond glasscutter. Handle the capillaries with grooved forceps, which have been flamed to ensure lack of RNAses.

2. For closing the small capillaries, melt the glass on one end using a gas burner.

3. In order to avoid any kind of contamination, flame the opened tip of the microtube.

4. Several microtubes can be filled with RNAs. The microtubes will be more easily handled by 'sticking' the closed end on Blue-tak or any clay-like substance, which is previously fixed on the cover of a 3.5 cm petri dish.

5. Transfer the RNA from the Eppendorf tube to the microtubes, using a MicroFil pipette (World Precision Instruments) adapted to a mouth pipette.

6. Put the filled capillaries into screw top Eppendorf tubes using sterile forceps.

7. Store at −80°C.

NOTES

Protocol 5.2: Microinjection of dsRNA into mouse oocytes and embryos

RECOVERY AND CULTURE OF OOCYTES AND EMBRYOS

Immature oocytes arrested at prophase I of meiosis are collected from ovaries of 4–6-week-old mice using a protocol described elsewhere.

To obtain the maximum numbers of fertilized eggs, female mice are superovulated by intraperitoneal injections of pregnant mare's serum gonadotrophin (PMSG, 5 IU), and human chorionic gonadotrophin (hCG, 5 IU) 48–52 h apart. Fertilized one-cell stage embryos are obtained from mated females 20–24 h after hCG injection. They are cleared of the cumulus cells with hyaluronidase treatment.

Oocytes and embryos are collected in FHM medium (*Table 5.2*) supplemented with BSA (1 mg/ml) (FractionV, Sigma) in plastic Petri

Table 5.2　Composition of FHM medium

	g for 100 ml
NaCl	0.55518
KCl	0.01864
KH_2PO_4	0.0476
$MgSO_4\,7H_2O$	0.049
Na lactate 60% syrup	0.15 ml
Glucose	0.0036
Na Pyruvate	0.0022
$NaHCO_3$	0.035
Glutamine	0.0146
EDTA	0.003
BSA	0.1
Hepes	0.49
Phenol Red	0.001
Penicillin	10 000 units
Streptomycin	10 mg
$CaCl_2\,2H_2O$	0.025

Preparation

Weigh out the Hepes and dissolve in 10 ml of H_2O. Adjust pH to 7.4. Weigh out the $CaCl_2$ and dissolve in 10 ml of H_2O. Weigh out the remaining components (except BSA and Lactate) and dissolve in 200 ml H_2O. Add Hepes, $CaCl_2$, and penicillin/streptomycin. Add lactate syrup, mix. Add BSA and allow to dissolve slowly. Mix and filter through Millipore filter (0.22 μm pore size). Store in aliquots at 4°C for 2 weeks or at −80°C for 3 months.

dishes (Falcon). For long time cultures, oocytes and embryos are transferred into pre-equilibrated drops of KSOM medium (Speciality Media Inc, Lavalette, NJ) supplemented with 4 mg/ml BSA, under oil in a humidified atmosphere of 5% CO_2 in air, at 37°C. Oil for culture is embryo-tested mineral oil (Sigma).

PREPARATION OF RNA FOR MICROINJECTIONS

Just before microinjection, the Eppendorf tubes containing the capillary tubes are centrifuged (12 000 *g*, at 4°C, 20 min to 1 h), in order to pellet any particles present in the RNA suspension that would block the injection needle.

PREPARATION OF INJECTION AND HOLDING PIPETTES

Classical methods are used to prepare injection and holding pipettes (see 'Manipulating the mouse embryo' for a detailed protocol).

Injection needles are made from borosilicate thin-wall capillaries – with filament, from Clark Electromedical Instruments. They have an outer diameter of 1.0 mm and inner of 0.78 mm. Injection needles are pulled on a model P-87 'Flaming/Brown Micropipette Puller' (Sutter Instruments, Inc.), although other types and models are commercially available.

Holding pipettes are made from capillaries which do not have an inside filament. The tip of holding pipettes has an external diameter of 80–100 μm, and an internal diameter of about 20 μm.

Micro-Fil pipettes connected to a mouth pipette are used to transfer the RNA from the microtube to the injection needle. Due to the small volume, it is necessary to transfer the RNA as close as possible to the injection tip; capillary action should pull the RNA into the very tip of the needle from there.

EQUIPMENT

In our laboratory, we use one of the several systems commonly used for microinjections into the oocyte or embryo, i.e. an inverted light microscope with Nomarsky differential interference contrast optics, which gives a very fine resolution of the oocyte and embryo structures. Other users prefer upright microscopes. With this type of microscope, hanging drops are commonly used because they usually give better optical results.

In order to improve penetrance of the plasma membrane, we set up an electrophysiological system utilizing negative capacitance. This allows the needle to pierce the membrane with minimal physical trauma. Once inside the membrane, we use a pressure injection system (Transjector System from Eppendorf) that delivers a constant flow of RNA, and allows more precise volumes to be delivered to the cell.

MICROINJECTIONS

Microinjections are carried out using glass depression slides. One drop of FHM medium is placed in the center of the depression and covered with embryo-tested mineral oil.

After microinjections, oocytes and embryos are transferred into drops of KSOM medium under oil, and incubated at 37°C (5% CO_2), where they will continue to develop.

ANALYSIS

If a fluorescent marker RNA (such as MmGFP) has been injected in conjunction with or by fusion to the RNA of interest, then fluorescence microscopy can be used to determine the success of the injection. The effects of dsRNA introduction will be deduced from alterations of development, and/or by analysis of the appropriate stage embryos for the expression of the targeted gene product (see above).

NOTES

Human embryonal carcinoma cells: surrogate tools in the study of human embryonic stem cells

6

Jonathan S. Draper and Peter W. Andrews

6.1 Introduction

The history of embryonic stem (ES) cells is intimately linked with the history and study of embryonal carcinoma (EC) cells derived from teratocarcinomas. Whether ES cells would have been discovered and studied in the same way without the previous studies of EC cells is a moot point, but it is clear that the earlier study of mouse EC cells (1,2) did provide many of the tools and concepts used in the derivation and study of ES cells, as well as the original rationale for their isolation (3,4).

Teratomas and teratocarcinomas are peculiar tumors that contain a wide range of haphazardly organized tissues (5). These tissues contain many of the cell types found in the adult as well as the cells found in extraembryonic membranes that envelop the developing fetus. Occasionally, the cells are organized into recognizable structures. Teeth and hair are often seen in ovarian cysts, the benign female manifestation of these tumors. More rarely, structures recognizable as limbs and other body parts may be found. Teratomas and teratocarcinomas are distinguished by the presence in teratocarcinomas of a small, primitive cell type, the embryonal carcinoma, or EC, cell, which appears to be the stem cell capable of differentiating into the various cell types that characterize these tumors as well as conferring malignancy. Teratomas without EC cells are commonly benign.

These tumors have fascinated medical scientists for many hundreds of years because they seem to represent a caricature of normal embryonic development (6). Before the advent of modern molecular genetics, and the discovery that the genetic systems that control development of model organisms, like the fruit fly or nematode worm, frequently have counterparts in mammalian embryos, the study of mammalian embryogenesis

Gene Targeting and Embryonic Stem Cells, Alison Thomson and Jim McWhir
© 2004 Garland Science/BIOS Scientific Publishers.

was particularly difficult. Mammalian embryos are typically small and inaccessible, and experimental embryology was mostly limited to studies of preimplantation embryos at the very early stage of their development. In this environment, teratomas and teratocarcinomas promised a route for investigating cell differentiation in the way that seemed pertinent to understanding the processes that occur in embryogenesis (7,8).

6.2   Embryonal carcinoma cells from the laboratory mouse

The experimental study of teratocarcinomas was initiated by Stevens who discovered that male mice of Strain 129 develop testicular teratomas and teratocarcinomas at an appreciable frequency (9). Stevens went on to show that these tumors could be derived in some other strains of mice by transplanting the genital ridge from mid-gestation embryos to an ectopic site. He also demonstrated that the tumors originated from the primordial germ cells shortly after they populated the genital ridge, between 11 and 13 days of development in the laboratory mouse (10,11). Later, it was shown that similar tumors could be derived from most strains of mice by transplanting embryos at the egg cylinder stage, about 7 days of gestation, to ectopic sites (12,13).

The ability of these tumors to be serially retransplanted to new host mice seemed to depend upon the presence of the EC cells. In a landmark study, Kleinsmith and Pierce were the first to show that, indeed, the transfer of a single EC cell to a new host mouse could lead to the formation of a histologically complex teratocarcinoma (14). Lines of EC cells from teratocarcinomas of the laboratory mouse were first established *in vitro* during the late 1960s in the laboratory of Boris Ephrussi (15). Such cells were then studied in detail by several groups (16–19).

Clonal lines of undifferentiated EC cells could be maintained, apparently indefinitely, in culture while retaining the capacity to differentiate when conditions were altered. Some lines required maintenance on feeder layers of mitotically inactivated mouse fibroblast lines, and would differentiate extensively when removed from the feeder cells. In particular, if these EC cells were prevented from attaching to a substrate, and maintained in suspension, they would form aggregates known as embryoid bodies in which, initially, a core of EC cells would be surrounded by cells resembling the primitive endoderm (17). Later these structures became cystic and histologically complex, and when allowed to attach, many differentiated cells would grow out from the clump (20). Other EC cell lines appeared not to require feeder cells for the retention of an undifferentiated phenotype, but they too would differentiate if allowed to grow to high densities (21). A significant advance was the discovery that differentiation could often be induced by exposure to particular chemical agents, notably retinoic acid (22,23), but also by agents like hexamethylene bisacetamide and dimethyl sulfoxide, known to induce differentiation of other stem cells, such as Friend erythroleukemia cells (24,25).

Comparison of EC cells with the early mouse embryo quickly led to the recognition that they closely resembled cells of the inner cell mass (ICM) of the blastocyst stage. Not only did EC and ICM cells share a similar morphology and developmental potential (in the mouse, both could evidently initiate differentiation leading to all somatic cell types, as well as extra-embryonic endoderm, but not trophectoderm), but they also shared expression of certain characteristic markers, notably the F9 antigen (26) and, later, the better-defined stage specific embryonic antigen-1 (SSEA-1) (27). The epitope identified with SSEA1 proved to be a carbohydrate with the Le^X structure carried on both glycolipids and glycoproteins (28–30).

These observations led to experiments to inject EC cells into blastocysts that were then replaced into the uterus of a pseudopregnant female mouse and allowed to develop to term. In a number of cases, the EC cells differentiated fully and their progeny became functionally incorporated into the developing embryo with the formation of a normal, chimeric adult mouse (31–33). Rarely, the germ line of these chimeras was reported to be populated by derivatives of the implanted EC cells (33,34). The results of these experiments clearly demonstrated the functional equivalence of EC cells from teratocarcinomas to ICM cells. They also highlighted the apparent regulation of the tumor phenotype of EC cells by the embryonic environment. Thus, whereas EC cells implanted into various sites in adult mice would form retransplantable teratocarcinomas, it initially appeared that this behavior was suppressed after transfer to a blastocyst.

However, it soon became apparent that this was often not the case and chimeras often go on to develop teratocarcinomas (35). This result is, perhaps, not surprising when it is remembered that EC cells have developed and adapted to tumor growth. Often, despite initial hopes to the contrary, many mouse EC cells proved to be not truly euploid but to carry small chromosomal rearrangements. Further, many have a reduced capacity for differentiation, or indeed do not seem to differentiate at all – the so-called 'nullipotent' lines. One might anticipate that EC cells would be subjected to strong selection pressure for the overgrowth of any genetic variants that do not differentiate, since their differentiated derivatives are typically not capable of prolonged proliferation, nor present a malignant phenotype. In support of this notion, hybrid cells formed between nullipotent EC cells and certain somatic cells often retain an EC phenotype (36), but they may also show more capacity to differentiate than the parental EC cell (37–39). Such a result would be consistent with the notion that recessive mutations that inhibit differentiation in the parent EC cells are complemented by wild-type alleles introduced by the somatic cell parent of such hybrids. Thus, in the phrase of Barry Pierce, EC cells are caricatures of normal pluripotent stem cells of the early embryo (40). While their similarities can be usefully exploited, not least because they can be grown in large amounts or because their differentiation can be controlled, it must always be remembered that they are cells that have adapted to growth conditions that differ from the early embryo.

Nevertheless, capitalizing on these studies of teratocarcinoma-derived EC cells, Evans and Kaufman (3), and separately Martin (4) demonstrated that it is possible to derive cell lines by explanting blastocysts onto feeder cells *in vitro*, under conditions used in some of the early derivations of EC cell lines. These embryo-derived cells closely resembled EC cells, expressing similar markers and differentiating into a range of cell types when removed from feeder cells *in vitro*, or inoculated into mice, when they formed teratomas. They soon became known as embryonic stem (ES) cells to distinguish them from EC cells. Like EC cells they also formed chimeras when replaced into a blastocyst that was allowed to develop to term. However, in this case, the ES cells proved to be substantially more efficient, often contributing to almost the whole embryo and, importantly, frequently contributing to the germ line. Evidently these cells much more closely resembled the pluripotent cells of the blastocyst, presumably because they had had less time to acquire genetic changes that adapted them to *in vitro* culture conditions. However, the ability to genetically manipulate these cells *in vitro*, particularly by selecting for homologous recombination has provided one of the key tools of modern mammalian experimental embryology, namely the production of transgenic mice in which the function of specific genes is modified (41).

With the advent of ES cell lines their principal application has been the production of transgenic mice, which permitted direct experimental analysis of murine embryogenesis, with relatively few studies focusing upon the biology of EC and ES cells themselves. However, the similarities of EC and ES cells to primordial germ cells (PGC) and the known PGC origin of testicular teratocarcinomas, led to further experiments to study this relationship. PGCs migrate into the genital ridge at about 11 days of gestation. When isolated shortly after this and cultured on feeder cells in the presence of leukemia inhibitory factor (LIF), stem cell factor (SCF) and basic FGF (fibroblast growth factor), these cells were found to proliferate and convert to cells that closely resembled EC and ES cells, expressing similar markers and developmental potential (42). To distinguish them they were known as EG cells. It seems possible that this conversion of PGC to EG cells in culture reflects some of the early events underlying the spontaneous formation of teratocarcinomas in the genital ridges of 129 mice. Nevertheless, despite study of the 129 mouse and teratocarcinomas for nearly 50 years, the underlying mechanism for this conversion and the genetic basis for its particular occurrence in the 129 Strain remains obscure.

On the other hand more progress has been made in understanding the mechanisms that regulate in ES cells the balance between self-renewal on the one hand and differentiation on the other. One aspect of this control depends upon expression of the transcription factor, Oct4; alterations in the levels of this factor have dramatic effects (43). Underexpression of Oct4 results in differentiation of the cells to yield trophectoderm, whereas overexpression also leads to differentiation but into extraembryonic endoderm. What controls the levels of Oct4 expression is unknown, but a separate level of regulation depends upon the interaction of ES cells with the cytokine, LIF (44). Whereas removal of ES cells from feeder cells typically leads to their differentiation, this can be prevented by exposure to LIF, so that it is now possible to maintain mouse ES cell lines in the absence of feeder cells (45). It appears that the function of LIF is mediated on the one hand by its activation of the STAT3 signaling pathway, and its inhibition of the Erk pathway on the other (46,47). How this *in vitro* phenomenon relates to the situation in the embryo *in vivo* is uncertain because *Lif-/Lif-* mouse embryos are able to develop (48). In fact, cells equivalent to ES cells do not normally persist long during normal embryonic development, and the function of LIF may relate to the situation of diapause, when mouse embryos can arrest development for significant periods, reinitiating embryo-genesis later when environmental conditions are more suitable for the survival of newborn mice (49).

6.3 Human embryonal carcinoma cells

While murine embryogenesis has now become relatively accessible to experimental study, human embryonic development remains largely inaccessible due to ethical as well as logistical constraints. Nevertheless, human and primate embryos differ from murine embryos in various respects and it might be assumed that a full understanding of human embryonic development cannot come from studies in the laboratory mouse alone. Therefore human EC cells derived from human teratocarcinomas might be expected to provide useful tools for investigating cell differentiation during human embryonic development, much as mouse EC and ES cells have provided tools for murine embryology (50). Following the studies of murine terato-carcinomas and the isolation and characterization of mouse EC cells, attempts were made to isolate EC cells from human teratocarcinomas, which are predominantly found as testicular cancers in young men. From a medical standpoint, testicular teratocarcinomas, which form a subset of so-called germ cell tumors, are of significant interest in their own right since they are the most common cancer of young men. Although they are also amongst the most curable solid tumors with current treatment, therapy is harsh, especially given the young age of the patients, and is still sometimes unsuccessful.

Initially, it was assumed that human EC cells would closely resemble mouse EC cells and in the first studies of cell lines derived from human testicular cancers, it was reported that the cells were SSEA1-positive, like murine EC cells (51,52). However, on more detailed examination, it appeared that, although cells closely resembling mouse EC cells in their general morphology and growth patterns were present in such cell lines, these cells did not express some of the markers characteristic of mouse EC cells, in particular they lacked SSEA1 (53). The morphology of a nullipotent human EC cell line, 2102Ep, is illustrated in *Figure 6.1A*. On the other hand, another antigen, SSEA3, which is expressed on the cleavage stage of mouse embryos but not on mouse ICM, EC or ES cells, was expressed strongly by these EC-like cells in both human teratocarcinoma-derived cell lines, and in the tumors themselves (54,55). Further studies of human tumors were extended with the discovery of another antigen, SSEA4, closely related to SSEA3, which was also expressed by human EC cells, but not mouse EC cells (56). SSEA3 and SSEA4, like SSEA1, proved to be associated with cell surface glycolipids, but they had a globoseries core structure instead of the lactoseries core that characterized SSEA1 (*Table 6.1*). SSEA1 does appear during the differentiation of human EC cells, apparently as a result of changes in synthesis of the respective core oligosaccharides (57,58).

Further similarities and differences are evident between the EC cells of both species. Thus, both mouse and human EC cells strongly express tissue nonspecific alkaline phosphatase (59) and they also express the transcription factor Oct4 (60; Draper and Andrews, unpublished results). On the other

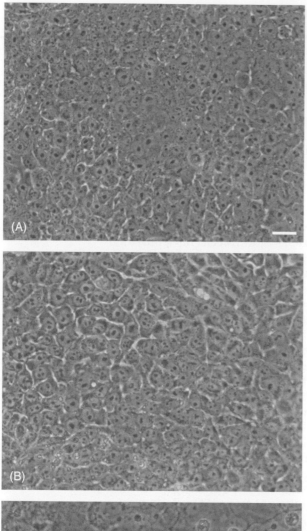

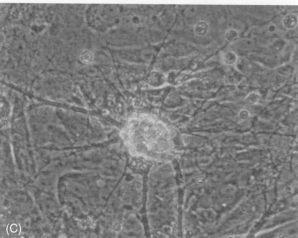

Figure 6.1

Human EC cells and neuronal derivatives. Human embryonal carcinoma cell lines 2102Ep (**A**) and NTERA2/D1 (**B**) are morphologically highly similar, processing multiple nucleoli and a high nuclear to cytoplasmic ratio. NTERA2/D1 can be induced to differentiate down neural lineage by the addition of retinoic acid, producing neurons (**C**) amongst other cell types. Scale bar = approx 70 μM.

Table 6.1 Glycolipid antigen markers of human EC and ES cell differentiation

Globoseries glycolipids	Antigen
Galβ1 → 3GalNAcβ1 → 3Galα1 → 3Galβ1 → 4Glcβ1 → Cer	SSEA3
NeuNAcα → 3Galβ1 → 3GalNAcβ1 → 3Galα1 → 3Galβ1 → 4Glcβ1 → Cer	SSEA3, SSEA4
Fucα1 → 2Galβ1 → 3GalNAcβ1 → 3Galα1 → 3Galβ1 → 4Glcβ1 → Cer	Globo H
GalNAcα1 → 3Galβ1 → 3GalNAcβ1 → 3Galα1 → 3Galβ1 → 4Glcβ1 → Cer	Globo A

$$2$$
$$\uparrow$$
$$\text{Fuc}\alpha1$$

Lactoseries glycolipids	Antigen
Galβ1 → 4GlcNAβ1 → Galβ1 → 4Glcβ1 → Cer	SSEA-1

$$3$$
$$\uparrow$$
$$\text{Fuc}\alpha1$$

Ganglioseries glycolipids	Antigen
(9-0-acetyl)NeuNAcα2 → 8NeuNAcα2 → 3Galβ1 → 4Glcβ1 → Cer	ME311
NeuNAcα2 → 8NeuNAcα2 → 8NeuNAcα2 → 3Galβ1 → 4Glcβ1 → Cer	A2B5

The globoseries antigens SSEA3, SSEA4, globo A and globo H (56,58) are all characteristic markers of human EC and ES cells. The lactoseries antigen SSEA1 (27,28) and the ganglioseries antigens, ME311 and A2B5 (58) are synthesised by extension of a common precursor, lactosyl ceramide and appear in some lineages upon differentiation.

hand, many human EC cells evidently have a capacity to differentiate into trophectoderm, as trophoblastic derivatives are common components of human teratocarcinomas and have been reported also in cultures of human EC cells (53,61). On the other hand, trophoblastic differentiation is not normally seen at all in murine EC or ES cells, and the embryonic cells to which they correspond, namely the cells of the late ICM or primitive ectoderm, have evidently lost the capacity to differentiate into trophectoderm. Whether human EC cells then did correspond to human ICM cells, or whether their embryological counterparts were distinct from those of mouse EC cells was a moot point.

A large number of human teratocarcinoma lines were established in culture but most appeared to consist of near nullipotent EC cells, perhaps because the likely long development period of these tumors in humans provided more opportunities for selection of variant cells that had lost the ability to differentiate (53,62,63). Indeed, human EC cells are highly aneuploid, typically with about 60 chromosomes, including many rearrangements. Nevertheless, these lines proved particularly valuable for exploring the general features of these cells, and further markers such as the antigens TRA-1-60 and TRA-1-81 and GCTM2 were identified after preparation and monitoring of antibodies from mice immunized with human EC cells (64,65). TRA-1-60, TRA-1-81 and GCTM2 are closely related members of a keratan sulfate family in which the carbohydrate side chains are modified to give a variety of determinants, although GCTM2 evidently is a determinant associated with the polypeptide backbone (66,67).

Several human EC cell lines were also isolated that retained an ability to differentiate. Ironically, the first to be studied in detail, TERA2, was one of the earliest to have been isolated (68), but difficulties in maintaining the cells in an undifferentiated state led to the recognition of its pluripotency being delayed (69). In nude (*nu/nu*) athymic mice, this line was shown to form tumors which contained a variety of tissues including neural and glandular structures as well as nodules of cartilage (70). From such a xenograft tumor, a subline, NTERA2, was derived and many studies have focused upon clonal sublines of this. In culture, NTERA2 cells could be maintained as undifferentiated EC cells if cultured at high cell densities, (*Figure 6.1B*). However, if exposed to retinoic acid they would differentiate losing all features of EC cells and converting to a variety of cells that included mature neurons (*Figure 6.1C*) (71,72). These neurons expressed ion channels and other features of mature neurons, and have become a widely used tool by neurobiologists to study properties of human fetal neurons (73–77). Moreover, they have been used in preliminary studies pertinent to the current interest in the use of stem cells in regenerative medicine. It has been reported that after transplantation into rat models of stroke NTERA2-derived neurons will allow partial recovery (78). The transplantation of these teratocarcinoma-derived neurons into human stroke patients has also been reported, but the efficacy of these trials remains to be seen (79).

NTERA2 cells respond to a variety of agents including hexamethylene bisacetamide (80) and bone morphogenetic proteins (81) which induce the differentiation into a variety of cell types, apparently separate from those induced by retinoic acid. They therefore provide useful tools for investigating the processes of cell differentiation in a way that should be pertinent to human development. Early studies, for example, demonstrated that differentiation of NTERA2 EC cells in response to retinoic acid resulted in the induction of the *Hox* genes in a retinoic acid dose-dependent manner that related to the position of the *Hox* genes in the various *Hox* gene clusters (82). Since this pattern also related to the pattern of *Hox* gene expression along the anterior/posterior axis of the developing embryo, it suggested a role for retinoic acid gradients in determining the anterior/posterior axis in early human, as well as mouse development. In other studies we showed that the sequence of neural-related gene expression during NTERA2 differentiation matched the temporal pattern of expression of these genes during neurogenesis in the early embryo (75).

Undifferentiated NTERA2 cells were also found not to be susceptible to infection with certain viruses such as human cytomegalovirus (HCMV) (83) and human immunodeficiency virus (HIV) (84), whereas their differentiated derivatives are permissive for the replication of these viruses. They have thus been used by a number of investigators interested in the mechanisms that regulate the replication of both viruses in embryonic tissues. In the case of HCMV this is of particular interest since HCMV is a significant cause of virally induced birth defects.

Several other human EC cell lines were also shown to be able to differentiate in culture. The most notable amongst these was GCT27, a human EC cell line that required maintenance on mitotically inactivated feeder cells (85–87). After removal from the feeder cells, these EC cells would differentiate giving rise to a variety of cell types. However, unlike feeder-dependent murine ES cells, their differentiation was not inhibited by LIF. Other human EC cell lines capable of differentiating include NCCIT (88,89) and NGR.R3 (90,91). These cells have been studied to a lesser extent than NTERA2, but they too are evidently capable of differentiating into a wide range of cell types.

6.4 Human embryonic stem cells

As in the case of the mouse, there was, early on, considerable interest in the possibility of deriving ES cells from human embryos that would parallel human EC cells. The different expression pattern of antigens and other markers by human EC cells compared to those of the mouse raised the question of whether human EC cells did correspond to pluripotent cells of the early human embryo, in the same way that mouse EC cells corresponded to the ICM of early mouse embryos. It would have been entirely possible that human EC cells corresponded to some different cell type during human embryogenesis. In any event, human EC cells were considerably more genetically abnormal than their murine counterparts, and their capacity for differentiation was often more restricted than that of the best mouse lines. Thus, their utility for studying embryonic cell differentiation was significantly less than for murine EC cells.

Although early human embryos became available for study following the development of *in vitro* fertilization (IVF) techniques during the 1970s, substantial logistical and ethical problems delayed progress in the isolation of human ES lines. However, primate ES cells were eventually derived, first from rhesus monkeys (92) and marmosets (93), and then from human embryos (94). Human ES cells were also derived from IVF embryos (95) and EG cells from human primordial germ cells (96). It was immediately evident that these primary ES and EG cells, which all required maintenance on feeder cells, closely resembled human EC cells in their morphology and general growth patterns, and in their expression of various key markers such as alkaline phosphatase and Oct4. In particular, human ES cells, like human EC cells, differed from murine EC and ES cells by expressing SSEA3, SSEA4, TRA-1-60 and TRA-1-81, but not SSEA1, which, nevertheless, is expressed by some of their differentiated derivatives (97). They also express, like human EC cells, the antigen Thy-1, whereas the mouse counterpart is not expressed by murine EC cells. Further, study of early human embryos showed that the ICM cells of human blastocysts also express SSEA3, SSEA4, TRA-1-60 and TRA-1-81, but not SSEA1 (98). Thus, human ES and EC cells do appear to correspond to the ICM of preimplantation human embryos, and the differences between mouse and human EC and ES cells seem to reflect differences between mouse and human embryos.

Other differences between mouse and human ES cells are evident. For example, human ES cells do not appear to respond to LIF, which is able to prevent differentiation of mouse, but not human, ES cells in the absence of feeder cells. At present, the significance of these various differences in the biology of human and mouse ES cells is unclear. Nevertheless, although there are some features in common, it is evident that one cannot necessarily extrapolate directly results obtained in the mouse to human. Studies of mouse embryos and mouse development can provide important information to guide studies of human development, but a full understanding of human embryonic development requires direct study with human cells. At the same time, human EC cells do continue to provide useful tools for investigating the biology of human ES cells, of which they present a caricature. Thus, whereas human ES cells remain difficult to maintain, the more robust character of established EC cell lines means that they can offer a useful surrogate for initial experiments, though results with EC cells should eventually be validated using ES cells themselves.

6.5   Surface antigens as tools for analysis of EC cells

Surface molecules recognized as antigens by specific monoclonal antibodies provide powerful tools for analyzing complex mixtures of cells such as those that occur in differentiating populations of cultured stem cells. By a variety of techniques, perhaps most powerfully by immunofluorescence using antibodies tagged with different fluorochromes, the expression of specific antigens can be detected on individual cells. Since this can be achieved using live cells, the expression of an antigen detected by an appropriate antibody can be used to isolate that cell for further functional analysis. Again, immunofluorescence coupled with flow cytometry and fluorescence-activated cell sorting (FACS) is perhaps the most powerful technique since it is readily adapted for the isolation of single cells and can be used to sort populations based on multiple parameters, such as expression of two or more antigens. However, other techniques such as the use of antibodies coupled

Table 6.2   Common antibodies used in FACS analysis

Antibody	Antigen	Antibody species and isotype	Reference
Markers of undifferentiated human EC and ES cells			
MC631	SSEA3	Rat IgM	54, 55, 56, 97, 98
MC813-70	SSEA4	Mouse IgG	56, 58, 97, 98
TRA-1-60	TRA-1-60	Mouse IgM	64, 67, 97, 98
TRA-1-81	TRA-1-81	Mouse IgM	64, 67, 97, 98
GCTM2	GCTM2	Mouse IgG	65, 66, 95
TRA-2-54	Liver-alkaline Phosphatase	Mouse IgG	97, 98, 102
TRA-2-49	Liver-alkaline Phosphatase	Mouse IgG	97, 98, 102
Cell surface antigen markers of differentiation			
MC480	SSEA1 (Le^x)	Mouse IgM	27, 58, 97, 98
A2B5	GT$_3$	Mouse IgM	58, 97, 100
ME311	9-0-acetylGD$_3$	Mouse IgG	58, 97, 101
VINIS56	GD$_3$	Mouse IgM	80, 97
VIN2PB22	GD$_2$	Mouse IgM	80, 97

to magnetic beads, panning and complement-mediated cytotoxicity can all also be used, though these are all more appropriate for isolation of populations of cells based upon a single parameter.

Human EC cells are characterized by the expression of the globoseries antigens SSEA3 and SSEA4, and the absence of the lactoseries antigen, SSEA1, which does appear upon their differentiation. Other key markers are the keratan sulfate antigens TRA-1-60, TRA-1-81 and GCTM2. On the other hand, differentiation can be monitored by the appearance of a variety of other antigens, apart from SSEA1. A number of these as well as the antibodies used to detect human EC cell-specific antigens are summarized in *Table 6.2*. Finally, we have found the pan human antigen, Ok(a), detected by antibody TRA-1-85 (99), useful in monitoring human EC and ES cells growing on mouse feeder cells. This antigen, originally identified as a red blood cell antigen, is expressed by all human cells we have analyzed, but it is not expressed by mouse cells; it can, therefore, be used to monitor the proportion of human cells, of whatever phenotype, in a mixed culture with mouse feeder cells (97).

6.5.1   Some words of caution

Surface antigen expression patterns are, operationally, very valuable markers of cell types within defined contexts. However, as with all markers, in the absence of other criteria it can be dangerous to claim identity for a particular cell outside the usual context in which the antigenic marker is commonly used. For example, the SSEA3 and SSEA4 antigens belong to the P blood group system and are found on red blood cells and a variety of other cell types. Within cultures of EC cells or in histological studies of biopsies of known germ cell tumors, they can be very valuable and seemingly specific for identifying EC cells. However, it would be foolhardy to use alone the expression of these antigens outside this context to claim the presence of EC or ES-like cells. The same caution must apply to all antigens used as cell markers, and the history of 'cell-specific antigen markers' is replete with examples of antigens first being claimed to be expressed uniquely by a particular cell type, only subsequently to be found expressed elsewhere on cells not examined in an initial screen. Another common fallacy is to assume that because two distinct cell types express a common antigen, that they therefore share a common lineage. Finally, caution must be exercised because many of the key markers used in studies of EC cells are carbohydrates. The function of these is often unclear and their expression, while seemingly robust in defined circumstances, may be affected by a variety of conditions that affect oligosaccharide synthesis and expression. Thus, we have noted sublines of the TERA2 EC cells that appear not to express the SSEA3 and SSEA4 determinants on their surfaces, although they do express the relevant glycolipids that carry these determinants (58). Further, although we believe that the immunodominant structure recognized on the cell surface by antibodies to SSEA3 and SSEA4 is the sialyated extended globoside structure, GL7 (*Table 6.1*) we routinely observe the disappearance of SSEA3 reactivity from differentiating EC and ES cells before the disappearance of SSEA4 reactivity – even though the antibodies apparent detect the same molecular entity (58).

6.5.2   Other markers: detection by *in situ* hybridization

A wide variety of molecular biology techniques, including northern analysis and RTPCR are routinely used to monitor the expression of particular genes that characterize the undifferentiated EC or ES cells, and their differentiated derivatives. The features of these techniques are not particularly different for this application than for any other. However, we have found *in situ* hybridization to be a useful complement to immunostaining for identifying subsets of stem and differentiated cells. We include a protocol that we have found of value.

Acknowledgments

This work was supported in part by grants from the Wellcome Trust and Yorkshire Cancer Research. We are grateful to Ian Morton for flow cytometry data.

References

1. Jacob F (1978) Mouse teratocarcinoma and mouse embryo. *Proc Roy Soc Lond B* 201: 249–270.
2. Solter D, Damjanov I (1979) Teratocarcinoma and the expression of onco-developmental genes. *Methods Cancer Res* **18**: 277–332.
3. Evans MJ, Kaufman MH (1981) Establishment in culture of pluripotential cells from mouse embryos. *Nature* **292**: 154–156.
4. Martin GR (1981) Isolation of a pluripotent cell line from early mouse embryos cultured in medium conditioned by teratocarcinoma stem cells. *Proc Natl Acad Sci USA* **78**: 7634–7636.
5. Damjanov I, Solter D (1974) Experimental teratoma. *Curr Topics Pathol* **59**: 69–130.
6. Wheeler JE (1983) History of teratomas. In: Damjanov I, Knowles BB, Solter D (eds) *The Human Teratomas: Experimental and Clinical Biology*, pp. 1–22. Humana Press, Clifton, NJ.
7. Martin GR (1980) Teratocarcinomas and mammalian embryogenesis. *Science* 209: 768–776.
8. Hogan BLM, Barlow DP, Tilly R (1983) F9 teratocarcinoma cells as a model for the differentiation of parietal and visceral endoderm in the mouse embryo. *Cancer Surv* **2**: 115–140.
9. Stevens LC, Little CC (1954) Spontaneous testicular teratomas in an inbred strain of mice. *Proc Natl Acad Sci USA* **40**: 1080–1087.
10. Stevens LC (1970) Experimental production of testicular teratomas in mice of strains 129, A/He and their F_1 hybrids. *J Natl Cancer Inst* **44**: 929–932.
11. Stevens LC (1964) Experimental production of testicular teratomas in mice. *Proc Natl Acad Sci USA* **52**: 654–661.
12. Solter D, Škreb N, Damjanov I (1970) Extrauterine growth of mouse egg-cylinders results in malignant teratoma. *Nature* **227**: 503–504.
13. Stevens LC (1970) The development of transplantable teratocarcinomas from intratesticular grafts of pre and post implantation mouse embryos. *Dev Biol* **21**: 364–382.
14. Kleinsmith LJ, Pierce GB (1964) Multipotentiality of single embryonal carcinoma cells. *Cancer Res* **24**: 1544–1552.
15. Kahn BW, Ephrussi B (1970) Developmental potentialities of clonal in vitro cultures of mouse testicular teratoma. *J Natl Cancer Inst* **44**: 1015–1029.
16. Martin GR, Evans MJ (1974) The morphology and growth of a pluripotent teratocarcinomas cell line and its derivatives in tissue culture. *Cell* **2**: 163–172.
17. Martin GR, Evans MJ (1975) Differentiation of clonal lines of teratocarcinomas cells: formation of embryoid bodies in vitro. *Proc Natl Acad Sci USA* **72**: 1441–1445.
18. Jakob H, Boon T, Gaillard J, Nicolas J-F, Jacob F (1973) Tératocarcinoma de la souris: isolement, culture, et proprieties de cellules à potentialities multiples. *Ann Microbiol Inst Pasteur* **124B**: 269–282.
19. Bernstine EG, Hooper ML, Grandchamp S, Ephrussi B (1973) Alkaline phosphatase activity in mouse teratoma. *Proc Natl Acad Sci USA* **70**: 3899–3903.
20. Martin GR, Wiley LM, Damjanov I (1977) The development of cystic embryoid bodies in vitro from clonal teratocarcinoma stem cells. *Dev Biol* **61**: 230–244.
21. Nicolas J-F, Dubois P, Jakob H, Gaillard J, Jacob F (1975) Tératocarcinome de la souris: différenciation en culture d'une lignée de cellules primitives à potentialities multiples. *Ann Microbiol Inst Pasteur* **126A**: 3–22.
22. Strickland S, Mahdavi V (1978) The induction of differentiation in teratocarcinomas stem cells by retinoic acid. *Cell* **15**: 393–403.

23. Strickland S, Smith KK, Marotti KR (1980) Hormonal induction of differentiation in teratocarcinoma stem cells: generation of parietal endoderm by retinoic acid and dibutyryl cAMP. *Cell* **21**: 347–355.

24. Jakob H, Dubois P, Eisen H, Jacob F (1978) Effets de l'hexaméthylènebis-acétamide sur la différenciation de cellules de carcinome embryonnaire. *CR Acad Sci (Paris)* **286**: 109–111.

25. McBurney MW, Jones-Villeneuve EM, Edwards MK, Anderson PJ (1982) Control of muscle and neuronal differentiation in a cultured embryonal carcinoma cell line. *Nature* **299**: 165–167.

26. Artzt K, Dubois P, Bennett D, Condamine H, Babinet C, Jacob F (1973) Surface antigens common to mouse cleavage embryos and primitive teratocarcinoma cells in culture. *Proc Natl Acad Sci USA* **70**: 2988–2992.

27. Solter D, Knowles BB (1978) Monoclonal antibody defining a stage-specific mouse embryonic antigen (SSEA-1). *Proc Natl Acad Sci USA* **75**: 5565–5569.

28. Gooi HC, Feizi T, Kapadia A, Knowles BB, Solter D, Evans MJ (1981) Stage-specific embryonic antigen involves α1-3 fucosylated type 2 blood group chains. *Nature* **292**: 156–158.

29. Childs RA, Pennington J, Uemura K, Scudder P, Goodfellow PN, Evans MJ, Feizi T (1983) High-molecular-weight glycoproteins are the major carriers of the carbohydrate differentiation antigens I, I and SSEA-1 of mouse terato-carcinoma cells. *Biochem J* **215**(3): 491–503.

30. Andrews PW, Knowles BB, Cossu G, Solter D (1982) Teratocarcinoma and mouse embryo cell surface antigens; Characterisation of the molecule(s) carrying the SSEA-1 antigenic determinant. In: Murumatsu T, Gachelin G, Moscona AA, Ikawa Y (eds) *Teratocarcinoma and Embryonic Cell Interactions*, pp. 103–119. Japan Scientific Societies Press, Tokyo.

31. Brinster RL (1974) The effect of cells transferred into the mouse blastocyst on subsequent development. *J Exp Med* **140**: 1049–1056.

32. Papaioannou VE, McBurney MW, Gardner RL, Evans MJ (1975) Fate of teratocarcinoma cells injected into early mouse embryos. *Nature* **258**: 70–73.

33. Mintz B, Illmensee K (1975) Normal genetically mosaic mice produced from malignant teratocarcinoma cells. *Proc Natl Acad Sci USA* **72**: 3585–3589.

34. Cronmiller C, Mintz B (1978) Karyotypic normalcy and quasi-normalcy of developmentally totipotent mouse teratocarcinoma cells. *Dev Biol* **67**: 465–477.

35. Rossant J, McBurney MW (1982) The developmental potential of a euploid male teratocarcinoma cell line after blastocyst injection. *J Embryol Exp Morphol* **70**: 99–121.

36. Miller RA, Ruddle FH (1976) Pluripotent teratocarcinomas – thymus somatic cell hybrids. *Cell* **9**: 45–55.

37. Andrews PW, Goodfellow PN (1980) Antigen expression by somatic cell hybrids of a murine embryonal carcinoma cell with thymocytes and L cells. *Somat Cell Genet* **6**: 271–284.

38. Rousset JP, Jami J, Dubois P, Aviles D, Ritz E (1980) Developmental potentialities and surface antigens of mouse teratocarcinoma x lymphoid cell hybrids. *Somatic Cell Genet* **6**: 419–433.

39. Rousset JP, Buchini D, Jami J (1983) Hybrids between F9 nullipotent terato-carcinoma and thymus cells produce multidifferentiated tumors in mice. *Devel Biol* **96**: 331–336.

40. Pierce GB (1974) Neoplasms, differentiations and mutations. *Am J Pathol* **77**: 103–118.

41. Thomas KR, Capecchi MR (1987) Site-directed mutagenesis by gene targeting in mouse embryo-derived stem cells. *Cell* **51**: 503–512.

42. Matsui Y, Zsebo K, Hogan BL (1992) Derivation of pluripotential embryonic stem cells from murine primordial germ cells in culture. *Cell* **70**: 841–847.

43. Niwa H, Miyazaki J, Smith AG (2000) Quantitiative expression of Oct-3/4 defines differentiation, dedifferentiation or self-renewal of ES cells. *Nature Genet* **24**: 372–376.
44. Smith AG, Heath JK, Donaldson DD, Wong GG, Moreau J, Stahl M, Rogers D (1988) Inhibition of pluripotential embryonic stem cell differentiation by purified polypeptides. *Nature* **336**: 688–690.
45. Pease S, Braghetta P, Gearing D, Grail D, Williams RL (1990) Isolation of embryonic stem (ES) cells in media supplemented with recombinant leukaemia inhibitory factor (LIF). *Dev Biol* **141**: 344–352.
46. Burdon T, Stracey C, Chambers I, Nichols J, Smith A (1999) Suppression of SHP-2 and ERK signalling promotes self-renewal of mouse embryonic stem cells. *Dev Biol* **210**: 30–43.
47. Niwa H, Burdon T, Chambers I, Smith A (1998) Self-renewal of pluripotent embryonic stem cells is mediated via activation of STAT3. *Genes Dev* **12**: 2048–2060.
48. Stewart CL, Kaspar P, Brunet LJ, Bhatt H, Gadi I, Kontgen F, Abbondanzo SJ (1992) Blastocyst implantation depends on maternal expression of leukaemia inhibitory factor. *Nature* **359**: 76–79.
49. Nichols J, Chambers I, Taga T, Smith A (2001) Physiological rationale for responsiveness of mouse embryonic stem cells to gp130 cytokines. *Development* **128**: 2333–2339.
50. Andrews PW, Goodfellow PN, Damjanov I (1983) Human teratocarcinoma cells in culture. *Cancer Surveys* **2**: 41–73.
51. Hogan B, Fellows M, Avner P, Jacob F (1977) Isolation of a human teratoma cell line which expresses F9 antigen. *Nature* **270**: 515–518.
52. Holden S, Bernard O, Artzt K, Whitmore WF Jr and Bennett D (1977) Human and mouse embryonal carcinoma cells in culture share an embryonic antigen (F9). *Nature* **270**: 518–520.
53. Andrews PW, Bronson DL, Benham F, Strickland S, Knowles BB (1980) A comparative study of eight cell lines derived from human testicular teratocarcinoma. *Int J Cancer* 26: 269–280.
54. Andrews PW, Goodfellow PN, Shevinsky L, Bronson DL, Knowles BB (1982) Cell surface antigens of a clonal human embryonal carcinoma cell line: Morphological and antigenic differentiation in culture. *Int J Cancer* **29**: 523–531.
55. Shevinsky L, Knowles BB, Damjanov I, Solter D (1982) Monoclonal antibody to murine embryos defines a stage-specific embryonic antigen expressed on mouse embryos and human teratocarcinoma cells. *Cell* **30**: 697–705.
56. Kannagi R, Cochran NA, Ishigami F, Hakomori S-i, Andrews PW, Knowles BB, Solter D (1983) Stage-specific embryonic antigens (SSEA-3 and -4) are epitopes of a unique globo-series ganglioside isolated from human teratocarcinoma cells. *EMBO J* **2**: 2355–2361.
57. Andrews PW, Knowles BB (1982) Human teratocarcinoma: Tools for human embryology. In: Murumatsu T, Gachelin G, Moscona AA, Ikawa Y (eds) *Teratocarcinoma and Embryonic Cell Interactions*, pp. 19–30. Japan Scientific Societies Press, Tokyo.
58. Fenderson BA, Andrews PW, Nudelman E, Clausen H, Hakomori S-i (1987) Glycolipid core structure switching from globo- to lacto- and ganglio-series during retinoic acid-induced differentiation of TERA 2 derived human embryonal carcinoma cells. *Dev Biol* **122**: 21–34.
59. Benham FJ, Andrews PW, Bronson DL, Knowles BB, Harris H (1981) Alkaline phosphatase isozymes as possible markers of differentiation in human teratocarcinoma cell lines. *Dev Biol* **88**: 279–287.
60. Yeom Y II, Fuhrmann G, Ovitt CE, Brehm A, Ohbo K, Gross M, Hubner K, Scholer HR (1996) Germline regulator element of Oct-4 specific for the totipotent cycle of embryonal cells. *Development* **122**: 881–894.

61. Damjanov I, Andrews PW (1983) Ultrastructural differentiation of a clonal human embryonal carcinoma cell line in vitro. *Cancer Res* **43**: 2190–2198.
62. Andrews PW, Damjanov I (1984) Cell lines from human germ cell tumours. In: Hay RJ, Park J-G, Gadzar A (eds) *Atlas of Human Tumour Cell Lines*, pp. 443–476. Academic Press, San Diego.
63. Andrews PW, Casper J, Damjanov I, *et al.* (1996) Comparative analysis of cell surface antigens expressed by cell lines derived from human germ cell tumors. *Int J Cancer* **66**: 806–816.
64. Andrews PW, Banting GS, Damjanov I, Arnaud D, Avner P (1984) Three monoclonal antibodies defining distinct differentiation antigens associated with different high molecular weight polypeptides on the surface of human embryonal carcinoma cells. *Hybridoma* **3**: 347–361.
65. Pera MF, Blasco-Lafita MJ, Cooper S, Mason M, Mills J, Monaghan P (1988) Analysis of cell differentiation lineage in human teratomas using new monoclonal antibodies to cytostructural antigens of embryonal carcinoma cells. *Differentiation* **39**: 139–149.
66. Cooper S, Pera MF, Bennett W, Finch JT (1992) A novel keratan sulphate proteoglycan from a human embryonal carcinoma cell line. *Biochem J* **286**: 959–966.
67. Badcock G, Pigott C, Goepel J, Andrews PW (1999) The human embryonal carcinoma marker antigen TRA-1-60 is a sialylated keratan sulphate proteoglycan. *Cancer Res* **59**: 4715–4719.
68. Fogh J, Trempe G (1975) New human tumor cell lines. In: Figh J (ed) *Human Tumor Cells in vitro*, pp. 115–159. Plenum Press, NY.
69. Andrews PW, Damjanov I, Simon D, Banting G, Carlin C, Dracopoli NC, Fogh J (1984) Pluripotent embryonal carcinoma clones derived from the human teratocarcinoma cell line Tera-2: Differentiation in vivo and in vitro. *Lab Invest* **50**: 147–162.
70. Duran C, Talley PJ, Walsh J, Pigott C, Morton I, Andrews PW (2001) Hybrids of pluripotent and nullipotent human embryonal carcinoma cells: partial retention of a pluripotent phenotype. *Int J Cancer* **93**: 324–332.
71. Andrews PW (1984) Retinoic acid induces neuronal differentiation of a cloned human embryonal carcinoma cell line in vitro. *Dev Biol* **103**: 285–293.
72. Lee VM-Y, Andrews PW (1986) Differentiation of NTERA-2 clonal human embryonal carcinoma cells into neurons involves the induction of all three neurofilament proteins. *J Neurosci* **6**: 514–521.
73. Rendt J, Erulkar S, Andrews PW (1989) Presumptive neurons derived by differentiation of a human embryonal carcinoma cell line exhibit tetrodotoxin-sensitive sodium currents and the capacity for regenerative responses. *Exp Cell Res* **180**: 580–584.
74. Squires PE, Wakeman JA, Chapman H, Kumpf S, Fiddock MD, Andrews PW, Dunne MJ (1996) Regulation of intracellular Ca2+ in response to muscarinic and glutamate receptor antagonists during the differentiation of NTERA2 human embryonal carcinoma cells into neurons. *Eur J Neurosci* **8**: 783–793.
75. Przyborski SA, Morton IE, Wood A, Andrews PW (2000) Developmental regulation of neurogenesis in the pluripotent human embryonal carcinoma cell line NTERA-2. *Eur J Neurosci* **12**: 3521–3528.
76. Pleasure SJ, Page C, Lee VM-Y (1992) Pure, post-mitotic, polarized human neurons derived from Ntera2 cells provide a system for expressing exogenous proteins in terminally differentiated neurons. *J Neurosci* **12**: 1802–1815.
77. Pleasure SJ, Lee VMY (1993) NTERA-2 cells a human cell line which displays characteristics expected of a human committed neuronal progenitor cell. *J Neurosci Res* **35**: 585–602.
78. Borlongan CV, Tajima Y, Trojanowski JQ, Lee VM, Sanberg PR (1998) Transplantation of cryopreserved human embryonal carcinoma-derived

neurons (NT2N cells) promotes functional recovery in ischemic rats. *Exp Neurol* 149: 310–321.

79. Kondziolka D, Wechsler L, Goldstein S, *et al.* (2000) Transplantation of cultured human neuronal cells for patients with stroke. *Neurology* **55**: 565–569.

80. Andrews PW, Nudelman E, Hakomori S-i, Fenderson BA (1990) Different patterns of glycolipid antigens are expressed following differentiation of TERA-2 human embryonal carcinoma cells induced by retinoic acid, hexamethylene bisacetamide (HMBA) or bromodeoxyuridine (BUdR). *Differentiation* **43**: 131–138.

81. Andrews PW, Damjanov I, Berends J, Kumpf S, Zappavingna V, Mavilio F, Sampath K (1994) Inhibition of proliferation and induction of differentiation of pluripotent human embryonal carcinoma cells by osteogenic protein-1 (or bone morphogenetic protein-7). *Lab Invest* **71**: 243–251.

82. Simeone A, Acampora D, Arcioni L, Andrews PW, Boncinelli E, Mavilio F (1990) Sequential activation of human HOX2 homeobox genes by retinoic acid in human embryonal carcinoma cells. *Nature* **346**: 763–766.

83. Gönczöl E, Andrews PW, Plotkin SA (1984) Cytomegalovirus replicates in differentiated but not undifferentiated human embryonal carcinoma cells. *Science* 224: 159–161.

84. Hirka G, Prakesh K, Kawashima H, Plotkin SA, Andrews PW, Gönczöl E (1991) Differentiation of human embryonal carcinoma cells induces human immunodeficiency virus permissiveness which is stimulated by human cytomegalovirus coinfection. *J Virol* **65**: 2732–2735.

85. Pera MF, Cooper S, Mills J, Parrington JM (1989) Isolation and characterization of a multipotent clone of human embryonal carcinoma cells. *Differentiation* **42**: 10–23.

86. Pera MF, Herzfeld D (1998) Differentiation of human pluripotent teratocarcinoma stem cells induced by bone morphogenetic protein-2. *Reprod Fertil Dev* **10**: 551–555.

87. Roach S, Schmid W, Pera MF (1994) Hepatocytic transcription factor expression in human; embryonal carcinoma and yolk sac carcinoma cell lines: Expression of HNF-3alpha in models of early endodermal cell differentiation. *Exp Cell Res* **215**: 189–198.

88. Teshima S, Shimosato Y, Hirohashi S, Tome Y, Hayashi I, Kanazawa H, Kakizoe T (1988) Four new human germ cell tumor cell lines. *Lab Invest* **59**: 328–336.

89. Damjanov I, Horvat B, Gibas Z (1993) Retinoic acid-induced differentiation of the developmentally pluripotent human germ cell tumor-derived cell line, NCCIT. *Lab Invest* **68**: 202–232.

90. Hata J, Fujita H, Ikeda E, Matsubayashi Y, Kokai Y, Fujimoto J (1989) Differentiation of human germ cell tumor cells. *Hum Cell* **2**: 382–387.

91. Umezawa A, Maruyama T, Inazawa J, Imai S, Takano T, Hata J (1996) Induction of mcl1/EAT, Bcl-2 related gene, by retinoic acid or heat shock in the human embryonal carcinoma cells, NCR-G3. *Cell Struct Funct* **21**: 143–150.

92. Thomson JA, Kalishman, J, Golos, TG, During M, Harris CP, Becker RA, Hearn JP (1995) Isolation of a primate embryonic stem cell line. *Proc Natl Acad Sci USA* **92**: 7844–7848.

93. Thomson JA, Kalishman J, Golos TG, During M, Harris CP, Hearn JP (1996) Pluripotent cell lines derived from the common marmoset (*Callithrix jacchus*) blastocysts. *Biol Reprod* **55**: 254–259.

94. Thomson JA, Itskovitz-Eldor J, Shapiro SS, Waknitz MA, Swiergiel JJ, Marshall VS, Jones JM (1998) Embryonic stem cell lines derived from human blastocysts. *Science* **282**: 1145–1147.

95. Reubinoff BE, Pera MF, Fong CY, Trounson A, Bongso A (2000) Embryonic stem cell lines from human blastocysts: somatic differentiation in vitro. *Nature Biotechnol* **18**: 399–404.

96. Shamblott MJ, Axelman J, Wang S, *et al.* (1998) Derivation of pluripotent stem cells from cultured human primordial germ cells. *Proc Natl Acad Sci USA* **95**: 13726–13731.

97. Draper JS, Pigott C, Thomson JA, Andrews PW (2002) Surface antigens of human embryonic stem cells: changes upon differentiation in culture. *J Anat* **200**: 249–258.

98. Henderson JK, Draper JS, Baillie HS, Fishel S, Thomson JA, Moore H, Andrews PW (2002) Preimplantation human embryos and embryonic stem cells show comparable expression of stage-specific embryonic antigens. *Stem Cells* **20**: 329–237.

99. Williams BP, Daniels GL, Pym B, Sheer D, Povey S, Okubo Y, Andrews PW, Goodfellow PN (1988) Biochemical and genetic analysis of the OKa blood group antigen. *Immunogenetics* **27**: 322–329.

100. Eisenbarth GS, Walsh FS, Nirenberg M (1979) Monoclonal antibody to a plasma membrane antigen of neurons. *Proc Natl Acad Sci USA* **76**: 4913–4917.

101. Thurin J, Herlyn M, Hindsgaul O, *et al.* (1985) Proton NMR and fast-atom bombardment mass spectrometry analysis of the melanoma-associated ganglioside 9-O-acetyl-GD3. *J Biol Chem* **260**: 14556–14563.

102. Andrews PW, Meyer LJ, Bednarz KL, Harris H (1984) Two monoclonal antibodies recognizing the determinants on human embryonal carcinoma cells react specifically with the liver isozyme of human alkaline phosphatase. *Hybridoma* **3**: 33–39.

Protocols

Contents

Protocol 6.1: Flow cytochemistry and fluorescence-activated cell sorting (FACS)

REAGENTS

0.25% trypsin (w/v): 1 mM EDTA in PBS (without Mg^{++} and Ca^{++})

DMEM/FCS (10% fetal calf serum in DMEM with 2 mM L-glutamine and antibiotics)

Wash buffer (5% fetal calf serum and 0.1% sodium azide in PBS)

EQUIPMENT

96 well plates

Orbital shaker

Cold room (at 4°C)

Centrifuge with microtitre plate carrier

Plate washer (with 12 position manifold)

Hemocytometer

FLOW CYTOMETRY

1. Harvest cells as a single cell solution by incubating with 1 ml 0.25% trypsin (w/v): 1 mM EDTA per T75 flask for 5 minutes at 37°C, inactivate the trypsin with 9 ml DMEM/FCS, count using a hemocytometer, centrifuge and resuspend at 2×10^6 cells per ml in wash buffer.

2. Dilute antibodies in wash buffer as appropriate and aliquot 50 µl of antibody per well of a 96-well plate. Add 50 µl of cell suspension (5×10^4 cells) per well and then incubate for 30 minutes at 4°C with gentle shaking.

3. Spin plates at 280 g for 3 minutes and dump the supernatant by quickly inverting the plate; the pellet should remain *in situ*.

4. Wash cells by adding 100 µl of wash buffer per well, centrifuge and dump supernatant. Repeat wash step two more times.

5. Resuspend cells in 50 µl of an appropriate dilution of FITC conjugated goat antimouse IgM or IgG (Ig fraction) secondary antibody (ICN) in wash buffer. Incubate for 30 minutes at 4°C with gentle shaking in the dark.

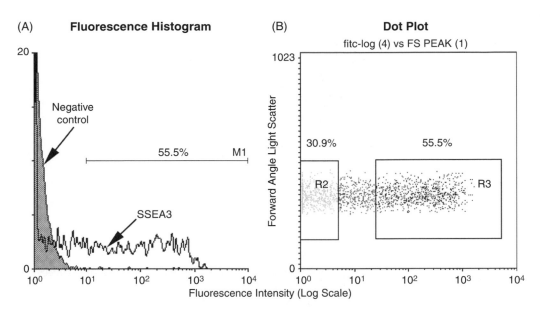

Analysis of Sorted Fractions

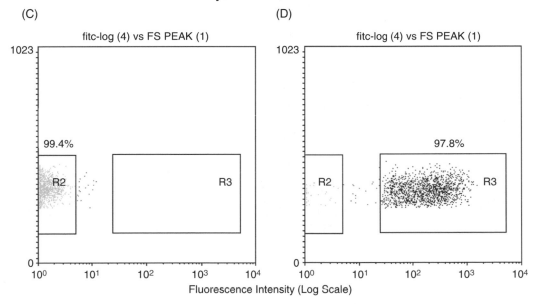

Figure 6.2

Flow cytofluorimetry and fluorescence-activated cell sorting of human ES cells. H7, human ES cells were stained for expression of SSEA3. (**A**) Fluorescence histogram showing the fluorescence distribution of cells stained with MC631 (anti-SSEA3) in comparison with staining with a negative control antibody. A total of 55.5% of the cells are judged to be SSEA3 positive. (**B**) Scatter plot showing fluorescence intensity against forward angle light scatter, a measure of cell size. The two regions shown designate the groups of cells that were separated as SSEA3 negative and positive populations. (**C, D**) Reanalysis of sorted SSEA3-negative and -positive subsets of cells.

6. Spin plates at 280 g for 3 minutes and dump the supernatant by quickly inverting the plate.

7. Wash cells by adding 100 μl of wash buffer per well, centrifuge and dump supernatant. Repeat wash step two more times.

8. Resuspend cells at 5 × 10^5 cells per ml in wash buffer, transfer to appropriate tubes and analyze.

RECLAMATION OF VIABLE CELLS USING FLUORESCENT-ACTIVATED CELL SORTING

1. Prepare primary and secondary antibodies by diluting as appropriate with wash buffer without sodium azide and filter sterilized using a 0.2 μm cellulose acetate filter.

2. Harvest-sufficient cells (at least 1 × 10^7) as a single cell solution by treatment with 1 ml of 0.25% trypsin (w/v): 1 mM EDTA per T75 flask for 5 minutes. Inactivate trypsin with 9 ml of DMEM/FCS per ml of trypsin used, transfer to a 15 ml screw cap tube, count using a hemocytometer and centrifuge at 188 g for 5 minutes.

3. Resuspend cell pellet with primary antibody at a ratio of 100 μl per 1 × 10^7 cells. Incubate, with gentle shaking, at 4°C for 20 minutes.

4. Wash cells by addition of DMEM/FCS to a total volume of 10 ml and centrifuge at 188 g at 4°C for 5 minutes, aspirate supernatant and repeat wash step.

5. Incubate cells with FITC conjugated goat anti mouse IgM or IgG (Ig fraction) secondary antibody (ICN) at 200 μl per 1 × 10^7 cells and incubate at 4°C for 20 minutes with gentle shaking.

6. Wash cells by addition of DMEM/FCS to a total volume of 10 ml and centrifuge at 188 g at 4°C for 5 minutes, aspirate supernatant and repeat wash step.

7. Resuspend cells in DMEM/FCS at 2 × 10^7 cells per ml and sort, collecting sorted cells into cold DMEM/FCS.

NOTES

Protocol 6.2: *In situ* immunofluorescence staining for surface antigens

REAGENTS

DMEM/FCS (10% fetal calf serum in DMEM with 2 mM L-glutamine and antibiotics)

PBS

4% PFA (4% paraformaldehyde in PBS)

EQUIPMENT

9 mm round glass coverslips

24-well plates

STAINING PROTOCOL

1. For visualization of cell surface antigens by *in situ* immunofluorescence cells do not require fixation.

2. Sterilize 9 mm round glass coverslips and place one per well of a 24-well plate.

3. Seed cells upon cover slips and grow cells as per normal until desired confluence is reached.

4. Aspirate medium and substitute with primary antibody diluted as appropriate in DMEM/FCS. Incubate in cell culture incubator for 30 minutes.

5. Aspirate medium and wash three times with DMEM/FCS.

6. Incubate in cell culture incubator with secondary antibody diluted as appropriate in DMEM/FCS for 30 minutes.

7. Wash once with DMEM/FCS and then twice more with PBS (containing Mg^{++} and Ca^{++}).

8. Cells can be visualized in the wells under PBS (containing Mg^{++} and Ca^{++}) or fixed with 4% PFA for 10 minutes and the individual cover slips mounted onto glass slides.

NOTES

Protocol 6.3: RNA *in situ* hybridization

REAGENTS

DEPC-PBS (PBS treated with 0.01% (v/v) solution of diethylpyrocarbonate)

4% DEPC-PFA (4% paraformaldehyde in DEPC-PBS)

Triethanolamine/acetic anhydride solution (25 ml 0.1 M triethanolamine, pH 8.0 and 62.5 μl acetic anhydride)

0.2 M HCl in DEPC treated dH_2O

20$\times$ SSC (3 M NaCl and 0.3 M sodium citrate dehydrate in dH_2O)

PBT (PBS + 0.1% Triton X-100)

Sheep serum (heat inactivated)

Anti-digoxygenin antibody (coupled to alkaline phosphatase)

Hybridization solution (For 100 ml: 50% formamide (Sigma), 5$\times$ SSC, 100 mg yeast tRNA, 10 mg heparin, 1$\times$ Denhardt's solution, 0.1% Tween 20, 0.1% CHAPS, 5 mM EDTA pH 8.0)

Alkaline phosphatase buffer (100 mM NaCl, 100 mM Tris HCl pH 9.5, 50 mM $MgCl_2$ and 0.1% Triton-X100)

Levamisole

EQUIPMENT

Tupperware box

Hybridization oven

IN SITU HYBRIDIZATION PROTOCOL

1. Fix cells in 4% DEPC-PFA at room temperature for 10 minutes. Wash three times in DEPC-PBS at room temperature for 5 minutes each.

2. Incubate at room temperature for 10 minutes in triethanolamine/acetic anhydride solution, followed by one wash in 1$\times$ SSC for 5 minutes at room temperature.

3. Treat with 0.2 M HCl in DEPC-water for 10 minutes, the cells were washed twice in DEPC-PBS for 5 minutes each.

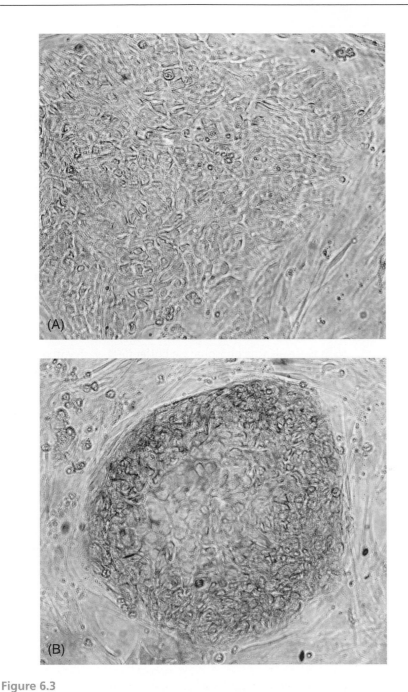

Figure 6.3

In situ hybridization for Oct-4 in human ES calls. Human ES cell colonies express the pluripotent stem cell marker Oct-4 as judged by *in situ* hybridization. Phase micrographs of hES colonies hybridized with (**A**) Oct-4 sense and (**B**) Oct-4 antisense probes.

4. Exchange the DEPC-PBS for hybridization solution and incubate for 6 hours at room temperature.

5. Replace with fresh hybridization solution and add an appropriate dilution of probe, typically between 1 and 2 μg/ml.

6. Hybridize overnight at 60–65°C (or as appropriate for probe) in a tight-sealing tupperware box containing towels soaked in 50% formamide and 5× SSC.

7. Rinse in 0.2× SSC and then wash in 0.2× SSC at 60 °C for 1 hour and allow to adjust to room temperature in 0.2× SSC for 5 minutes.

8. Block with in 20% sheep serum in PBT for at least 1 hour at room temperature.

9. Incubated with antidigoxygenin antibody diluted to a final concentration of 1:1000 in 20% sheep serum in PBT at 4°C overnight.

10. Rinse three times and then wash four times for 5 minutes each in PBT.

11. Wash twice alkaline phosphatase buffer (first wash without levamisole, second wash with one drop levamisole per 5 ml alkaline phosphatase buffer).

12. Visualize dioxygenin labeling by incubating with alkaline phosphatase buffer with 4.5 μl/ml of NBT (Roche) and 3.5 μl/ml of BCIP (Roche), and develop in the dark for between 2–36 hours, depending on the abundance of the RNA.

13. After the reaction has proceeded as desired, wash cells in PBT, and fix in 4% PFA.

NOTES

Isolation of human embryonic stem cells

7

L. Ahrlund-Richter and R. Pedersen

7.1 Introduction

In mammalian development, cells of the inner cell mass (ICM) of the blasto-cyst are destined to develop into the embryo proper. Remarkably, under certain conditions cells from the ICM can be adapted to growth *in vitro* and still retain developmental capacity. Such cells are likely to be the most developmentally flexible of all cultured cells and have been termed embry-onic stem (ES) cells, distinguishing them from other pluripotent cell lines, such as embryonal carcinoma (EC) and embryonic germ (EG) cell lines.

Pluripotent ES cells were first demonstrated in the mouse [1,2] and pro-vided important new models for studies on differentiation *in vitro*, extending the preceding models of EC cell lines. When combined with molecular biology, the early findings in the mouse led to the development of a whole animal approach to mammalian functional genomics involving 'gene targeting' *in vivo* (for review see [3]).

ES lines have been derived from primates, including rhesus monkey [4], marmoset [5] and human [6,7]. Bongso *et al.* [8] and Lavoir *et al.* [9] reported on short-term cultures from the inner cell mass of human blastocysts, but their colonies had a brief lifetime due to differentiation during early pas-sages. The experimental protocol used in the successful primate derivation studies was very similar to the ones previously used for mouse (for review see [10,11]).

What is perhaps surprising, in view of the successful derivation of pri-mate embryonic stem cells, is the documented difficulty in deriving pluripo-tent stem cell lines from embryos of domestic species. Derivation attempts using similar protocols for bovine embryos [12] and ovine embryos [13,14] have yielded cells with a tendency to differentiate rather than maintain their pluripotency. Pluripotent stem cells derived from chicken embryos have shown the ability to contribute to feather pigmentation and occasionally, to the germ line of chimeras [15]. Vertebrate embryo-derived pluripotent stem cell lines of hamster [16], mink [17], pig [18] and recently, rats [19] have shared certain features of mouse embryonic stem cells, but importantly have not contributed to the germ line of chimeras (for review see [20]). Thus, the fact that stable stem cell lines with demon-strable pluripotency can be derived from preimplantation primate embryos is fortuitous and has important implications both for therapeutic applica-tions and as a model system for understanding mechanisms of primate development at early stages not otherwise accessible for analysis [21].

Gene Targeting and Embryonic Stem Cells, Alison Thomson and Jim McWhir
© 2004 Garland Science/BIOS Scientific Publishers.

While conceptually straightforward, the derivation of pluripotent stem cells from human preimplantation embryos requires an exacting approach, as described here.

7.2 Derivation of hES cells

The technical process of deriving human embryonic stem cells from the inner cell mass of the blastocyst can be divided into three parts: (a) obtaining suitable embryos; (b) isolating the inner cell mass from the blastocyst; and (c) propagating the inner cell mass as pluripotent stem cells.

7.2.1 Obtaining suitable embryos

A central requirement in the process of deriving ES cell lines from the inner cell mass of human blastocysts is obviously the acquisition of embryos. For this, two approaches have been reported. The first is the use of cleavage-stage human embryos, produced by *in vitro* fertilization for clinical purposes. After informed consent, surplus embryos from *in vitro* fertilization programs were donated by patient couples (6–8,22–25).

Another approach involves the use of surplus embryos that following preimplantation genetic diagnosis have been identified with a high risk of genetic disease and are thus not suitable for embryo transfer (26). Alternatively, embryos generated for research could be used (27). Both approaches allow embryos of higher quality to be used for research since in clinical programs the best embryos would always be reserved for patient treatment. However, the latter approach raises the additional ethical issue involved in generating human embryos solely for research purposes.

A technically more advanced option would be the production of pluripotent cells by somatic cell nuclear transfer (28; Lavoir M-C, Fung JL, Conaghan J, Pedersen RA, unpublished observations). While feasible in generating mouse ES cell lines, this approach has not yet proved effective for generation of blastocysts from somatic cell nuclear transfer to human oocytes (see (29) for review). In fact, recent data from nonhuman primates suggest that changes in the current experimental protocols may be necessary in order to achieve viability of the primate embryo generated by enucleation and nuclear transfer (30). An alternative approach involving transfer of human somatic nuclei to rabbit oocytes reportedly yields blastocysts that are capable of giving rise to proliferating cultures resembling embryonic stem cells (31). It remains to be determined, however, whether such cultures are capable of normal nuclear–mitochondrial interactions that would be necessary to sustain their long-term growth (32).

7.2.2 Isolation of the ICM

Several methods for isolation of the ICM have been reported. In the most straightforward approach the ICM is mechanically separated from the trophoblastic cells using glass pipettes or by laser. Empirically, both these approaches demand that the person performing the isolation is able to clearly distinguish the trophoblastic cells from the inner cell mass. This is not easy and for many of the blastocysts available for this type of research, the inner cell mass is highly distended, and thus is not easy to detect by eye. Accordingly, separation of the different cell types this way is time consuming and not always effective. An alternative method to obtain pure inner cell masses in a relatively short time period is to use an immunosurgical method originally described for isolation of inner cell masses of the mouse blastocyst (33). In immunosurgery blastocysts are first exposed to a heat-inactivated species-specific antiserum, which binds to the outer trophoblast surface. Excess antibody is washed off, and the embryo is exposed to active complement. This results in the selective death of trophoblastic cells and recovery of the inner cell mass separate from the remnants of trophoblastic cells. The majority of the human ES cell lines available today have been derived using this method.

7.2.3 Propagation of ICM as pluripotent stem cells

The isolated ICMs are subsequently plated onto a supportive matrix, such as mitotically inactivated feeder cell monolayers and allowed to proliferate. Published reports describe successful derivations on mouse embryonic fibroblasts (6,7,24,26,27), the mouse embryonic cell line STO (25), or on human fibroblast cells derived from human fetal muscle (22) or human foreskin (23).

Initial outgrowth during the first week is often not easy to observe but contrary to the mouse derivation, it is often relatively easy to distinguish the individual cells in a good colony of human ES cells at early stages of derivation, as described in the accompanying protocols. It is essential to recognize the correct cell type and handle it appropriately during subculture. The doubling time varies greatly between different cultures and ranges from massive apoptosis to rapidly dividing and quickly differentiating cells. Successful colonies have been reported to have a population doubling time of about 36 hours (23,34). The initial colonies are sometimes raised (i.e. somewhat three-dimensional) but with time the colonies stretch out into a relatively flat appearance. Colonies of appropriate undifferentiated morphology are subsequently selected and expanded as described in the Protocols.

Procedures regarding further optimal growth and expansion will be described in Chapter 8.

7.3 Towards further optimized conditions

Published protocols for the derivation of human ES cells are similar but not identical. In line with this, the reported success rate differs. The status of the blastocyst is often poorly described and includes only rarely a standardized scoring of the donated embryo or details on its *in vitro* development. This makes any comparison on efficacy of the different protocols difficult, if not impossible. The use of two-stage *in vitro* culture systems for the blastocysts, that employ different media for appropriate developmental stages (reviewed by (35)), has greatly increased the quality of the blastocysts available also for

Table 7.1 Steps towards optimized derivation

- Find alternatives to the use of antiserum and complement from other species for the isolation of ICM
- Derivation using defined media, free of *all* xeno-derived components
- Use of defined matrix as growth support (preferably of recombinant source) instead of feeder cells

stem cell research. A well-developed ICM is for obvious reasons likely to contribute to success.

Derivation of pluripotent stem cell lines using feeder-free conditions has not yet been published, but would seem to be feasible given that established lines are amenable to culture under feeder-free conditions (36,38). For several reasons it is desirable to develop protocols avoiding the use of components that are poorly defined (such as sera) or components that are not possible to fully standardize (such as feeder cells). For the prospective future use in clinical protocols it will be highly desirable to avoid all xeno-derived components. *Table 7.1* summarizes the most essential steps towards optimized conditions and the possibility for derivations under conditions meeting *Good Manufactory Production* standards.

References

1. Evans MJ, Kaufman MH (1981) Establishment in culture of pluripotential cells from mouse embryos. *Nature* **292**: 154–156.
2. Martin GR (1981) Isolation of a pluripotent cell line from early mouse embryos cultured in medium conditioned by teratocarcinoma stem cells. *Proc Natl Acad Sci USA* **78**: 7634–7638.
3. Joyner AL (2000) *Gene Targeting: A Practical Approach*, 2nd edn. Oxford University Press, Oxford.
4. Thomson JA, Kalishman J, Golos TG, Durning M, Harris CP, Becker RA, Hearn JP (1995) Isolation of a primate embryonic stem cell line. *Proc Natl Acad Sci USA* **92**: 7844–7848.
5. Thomson JA, Kalishman J, Golos TG, Durning M, Harris CP, Hearn JP (1996) Pluripotent cell lines derived from common marmoset (*Callithrix jacchus*) blastocysts. *Biol Reprod* **55**: 254–259.
6. Thomson JA, Itskovitz-Eldor J, Shapiro SS, Waknitz MA, Swiergiel JJ, Marshall VS, Jones JM (1998) Embryonic stem cell lines derived from human blastocysts. *Science* **282**: 1145–1147.
7. Reubinoff BE, Pera MF, Fong CY, Trounson A, Bongso A (2000) Embryonic stem cell lines from human blastocysts: somatic differentiation in vitro. *Nat Biotechnol* **18**: 399–404.
8. Bongso TA, Fong CY, Ng CY, Ratnam SS (1994) Blastocyst transfer in human in vitro fertilization: the use of embryo co-culture. *Cell Biol Int* **18**: 1181–1189.
9. Lavoir MC, Conaghan J, Pedersen RA (1998) Culture of human embryos for studies on the derivation of human pluripotent cells: a preliminary investigation. *Reprod Fertil Dev* **10**(7–8): 557–561.
10. Robertson EJ (1987) Embryo-derived stem cell lines. In: Robertson EJ (ed) *Teratocarcinomas and Embryonic Stem Cells: A Practical Approach*, pp. 71–112. IRL Press, Oxford.
11. Marshall VS, Waknitz MA, Thomson JA (2001) Isolation and maintenance of primate embryonic stem cells. *Meth Mol Biol* **158**: 11–18.
12. First NL, Sims MM, Park SP, Kent-First MJ (1994) Systems for production of calves from cultured bovine embryonic cells. *Reprod Fertil Dev* **6**(5): 553–562.
13. Campbell KH, McWhir J, Ritchie WA, Wilmut I (1996) Sheep cloned by nuclear transfer from a cultured cell line. *Nature* **380**(6569): 64–66.
14. Talbot NC, Rexroad CE Jr, Pursel VG, Powell AM (1993) Alkaline phosphatase staining of pig and sheep epiblast cells in culture. *Mol Reprod Dev* **36**(2): 139–147.
15. Pain B, Clark ME, Shen M, Nakazawa H, Sakurai M, Samarut J, Etches RJ (1996) Long-term in vitro culture and characterisation of avian embryonic stem cells with multiple morphogenetic potentialities. *Development* **122**(8): 2339–2348.
16. Doetschman T, Williams P, Maeda N (1988) Establishment of hamster blastocyst-derived embryonic stem (ES) cells. *Dev Biol* **127**(1): 224–227.
17. Sukoyan MA, Vatolin SY, Golubitsa AN, Zhelezova AI, Semenova LA, Serov OL (1993) Embryonic stem cells derived from morulae, inner cell mass, and blastocysts of mink: comparisons of their pluripotencies. *Mol Reprod Dev* **36**(2): 148–158.
18. Wheeler MB (1994) Development and validation of swine embryonic stem cells: a review. *Reprod Fertil Dev* **6**(5): 563–568.
19. Fandrich F, Lin X, Chai GX, *et al.* (2002) Preimplantation-stage stem cells induce long-term allogeneic graft acceptance without supplementary host conditioning. *Nature Med* **8**(2): 107–108.
20. Prelle K, Zink N, Wolf E (2002) Pluripotent stem cells – model of embryonic development, tool for gene targeting, and basis of cell therapy. *Anat Histol Embryol* **31**(3): 169–186.

21. Pedersen RA (1999) Embryonic stem cells for medicine. *Sci Am* **280**(4): 68–73.
22. Richards M, Fong CY, Chan WK, Wong PC, Bongso A (2002) Human feeders support prolonged undifferentiated growth of human inner cell masses and embryonic stem cells. *Nat Biotechnol* **20**(9): 933–936.
23. Hovatta O, Mikkola M, Gertow K, *et al.* (2003) A culture system using human foreskin fibroblasts as feeder cells allows production of human embryonic stem cells. *Human Reprod* **18**(7): 1404–1409.
24. Mitalipova M, Calhoun J, Shin S, *et al.* (2003) Human embryonic stem cell lines derived from discarded embryos. *Stem Cells* **21**(5): 521–526.
25. Park JH, Kim SJ, Oh EJ, Moon SY, Roh SI, Kim CG, Yoon HS (2003) Establishment and maintenance of human embryonic stem cells on STO, a permanently growing cell line. *Biol Reprod* **69**(6): 2007–2014.
26. Pickering SJ, Braude PR, Patel M, Burns CJ, Trussler J, Bolton V, Minger S (2003) Preimplantation genetic diagnosis as a novel source of embryos for stem cell research. *Reprod Biomed Online* **7**(3): 353–364.
27. Lanzendorf SE, Boyd CA, Wright DL, Muasher S, Oehninger, Hodgen GD (2001) Use of human gametes obtained from anonymous donors for the production of human embryonic stem cell lines. *Fertility Sterility* **76**(1): 132–137.
28. Cibelli JB, Kiessling AA, Cunniff K, Richards C, Lanza RP, West MD (2001) E-biomed. *J Regener Med* **2**: 25.
29. Bradley JA, Bolton EM, Pedersen RA (2002) Stem cell medicine encounters the immune system. *Nat Rev Immunol* **2**(11): 859–871.
30. Simerly C, Dominko T, Navara C, *et al.* (2003) Molecular correlates of primate. Nuclear transfer failures. *Science* **300**: 297.
31. Chen Y, He ZX, Liu A, *et al.* (2003) Embryonic stem cells generated by nuclear transfer of human somatic nuclei into rabbit oocytes. *Cell Res* **13**(4): 251–263.
32. Kenyon L, Moraes CT (1997) Expanding the functional human mitochondrial DNA database by the establishment of primate xenomitochondrial cybrids. *Proc Natl Acad Sci USA* **94**(17): 9131–9135.
33. Solter D, Knowles BB (1975) Immunosurgery of mouse blastocyst. *Proc Natl Acad Sci USA* **72**(12): 5099–5102.
34. Amit M, Carpenter MK, Inokuma MS, *et al.* (2000) Clonally derived human embryonic stem cell lines maintain pluripotency and proliferative potential for prolonged periods of culture. *Dev Biol* **227**: 271–278.
35. Gardner DK (1998) Development of serum-free medium for the culture and transfer of human blastocysts. *Human Reprod* **13**(Suppl. 4): 218–225.
36. Xu C, Inokuma MS, Denham J, Golds K, Kundu P, Gold JD, Carpenter MK (2001) Feeder-free growth of undifferentiated human embryonic stem cells. *Nature Biotechnol* **19**(10): 971–974.
37. Amit M, Margulets V, Segev H, Shariki K, Laevsky I, Coleman R, Itskovitz-Eldor J (2003) Human feeder layers for human embryonic stem cells. *Biol Reprod* **68**(6): 2150–2156.
38. Amit M, Shariki C, Margulets V, Itskovitz-Eldor J (2004) Feeder and serum-free culture of human embryonic stem cells. *Biol Reprod* **70**(3): 837–845.
39. Abbondanzo SJ, Gadi I, Stewart CL (1993) Derivation of embryonic stem cell lines. *Methods Enzym* **225**: 803–823.

Protocol

Contents

Protocol 7.1: Immunosurgery: Separation of inner cells mass (ICM) from trophectoderm cells (TE)

(After (33); with minor modifications by J. Inzunza and M. Andäng, unpublished).

MATERIALS

Pronase (Sigma: P5147)

Antihuman whole serum (Sigma; H3383)

Guinea pig complement serum (Sigma: S1639)

Blastocyst medium (r S-2, or IVF-20; Vitrolife)

hES medium (see below)

Glass capillary micropipettes are made from a 1.5 mm hard glass capillary tube (BDH Laboratory Supplies, England) and pulled to create a very fine edge. The edge is broken under a stereo microscope to get an inside diameter of 0.040–0.050 mm. The micropipettes are sterilized at 180°C for 3 hours before use.

METHODS

Using a transfer glass capillary pipette the blastocyst is transferred into a drop of Pronase (0.5%; dissolve 0.025 g in 5 ml blastocyst medium) and incubated at 37°C until the zona pellucida is dissolved (1–2 minutes). Blastocysts without zona pellucida are immediately transferred into a new drop of blastocyst medium and washed 3× by transfer to new drops of blastocysts medium.

The zona free blastocyst is then:

1. incubated for 30 minutes at 37°C with antihuman antibody, diluted in blastocyst medium (1:5 dilution)

2. washed 4× in drops of blastocyst medium at 37°C

3. further incubated 20–30 min at 37°C in a drop of guinea pig complement (diluted 1:5) with blastocyst medium. The swelling process of the dying trophoblast cells is followed in an inverted

microscope and the process stopped before the ICM cells are affected

4. washed three times in drops of blastocyst medium at 37°C

Incubate the ICM in a drop of hES medium at 37°C, until the dead trophoblast cells are dislodged from the ICM. This is facilitated by gently pipetting the blastocysts cell mass through a glass capillary micropipette.

The ICM is subsequently placed into a tissue culture dish containing prewarmed hES culture medium and feeder cells. Great care is taken to avoid any TE contamination when the ICM is picked up from the dissection drop. It is advisable to mark the position of the clump in the culture dish.

CULTURE CONDITIONS

Primary mouse embryonic fibroblasts (MEF)

The derivation, passaging and freezing of mouse embryonic fibroblasts (MEF) have been reported extensively and a full description can be found, for example in (39).

Procedure notes

The influence of the donor mouse genotype is not fully known, but CF-1 or C57/Bl6 mice have been found to be good choices. The derivation of MEF is best performed using medium without antibiotics, and it is vital to test each new batch of MEF for microbial contamination, especially for subclinical infections such as mycoplasma. It is important to use mice of defined microbial status as the source of mouse fetuses.

The MEF cells can be mitotically inactivated by gamma irradiation or Mitomycin C treatment (39). The choice between using Mitomycin C or gamma irradiation for the inactivation of feeder cells seems mostly to be a matter of availability of a good irradiation source. It is preferable not to irradiate MEFs until immediately prior to use. The inactivated MEFs are useful as feeders for up to 2 weeks. Plate well suspended, low passage (passage 3–7) MEFs on gelatin-coated dishes (0.1% Gelatin (Sigma, G1890)).

Human foreskin fibroblasts

The use of human foreskin fibroblast is described by (23) and (38) Commercially available human foreskin fibroblasts (CRL-2429) can be obtained from ATCC Mananas, USA.

Procedure notes

The use of primary human foreskin fibroblasts is similar to the protocol for MEF. Early passages are preferable, but human foreskin fibroblasts have been found to effectively support hES cell growth using the fibroblast cells up to at least passage 16. The fibroblasts are grown to form a confluent monolayer, iradiated (35 Gy) and plated on the culture dish to be used as growth support. Suggested concentration for a 4-well plate is 2×10^4 cells/well and for a 6-well plate: 2×10^5 cells/well. The inactivated foreskin fibroblasts are useful as feeders for up to 2 weeks, possibly longer.

Published culture medium for ICM/ early hES cells

Dulbecco's modified Eagle's medium (DMEM); no pyruvate, high glucose formulation.

(6, 23, 24)	80% vol/vol (Gibco-BRL)
(7, 25, 26)	80% vol/vol (Life Technologies)
(27)	80% vol/vol (Speciality Media)
(22)	80% vol/vol (InVitrogen)

Fetal bovine serum[1]

(6, 7, 26, 24)	20% vol/vol (Hyclone)
(27)	20% vol/vol (Speciality media)
(23)	20% vol/vol (R&D Sweden)
(25)	20% vol/vol (supplier not reported)

Human serum

(22)	20% vol/vol (supplier not reported)

L-Glutamine

(6, 25)	1 mM (Sigma)
(27)	1 mM (Speciality Media)
(22)	'1×' (InVitrogen)
(7, 26)	2 mM (Life Technologies)
(23)	2 nmol/l (R&D Sweden)
(24)	2 mM (Gibco-BRL)

β-Mercaptoethanol

(6, 25)	0.1 mM (Sigma)
(7, 26)	0.1 mM (Life Technologies)
(27)	0.1 mM (Speciality Media)
(23, 24)	0.1 mM (Gibco-BRL)
(22)	1 mM (Sigma)

---

[1] The capacity of various batches of FBS to adequately support pluripotent stem cell growth varies tremendously and it is thus vital to select a good batch. The criteria for selecting the sera are good proliferation and induction of a minimum amount of differentiation, i.e. similar criteria to selection of FBS for mouse ES cell cultures for the generation of knockout mice (39).

Nonessential amino acids

(6, 23, 24)	'1%' (Gibco-BRL)
(7, 26, 25)	'1%' (Life Technologies)
(27)	'1%' (Speciality Media)
(22)	'1×' (InVitrogen)

Human recombinant leukemia inhibitory factor (hLIF);

(7, 26)	2,000 units/ml (Amrad)
(27)	0.1 μg/ml (Sigma)
(23)	1 μl/ml (Chemicon)
(25)	2000 units/ml (supplier not reported)
(24)	1000 units/ml (Chemicon)

Human insulin-transferrin-selenium

(24)	'1×' (InVitrogen)

Basic fibroblast growth factor (bFGF)

(24)	4 ng/ml (Gibco-BRL)

Addition of penicillin + streptomycin was used in studies 7, 22, 23, 26, 25, 24.

The complete culture medium is filter sterilized using a 0.22 μm filter and stored at +4°C for maximum 1 week.

PROCEDURES FOR EARLY PASSAGES

It is essential to observe the culture in an inverted microscope daily. Previous studies seem all to verify the following general growth behavior of the early passages: Groups of small, tightly packed cells are observed after 4–19 days.[2] The first colony can mechanically be dissociated into smaller clumps by the use of a micropipette, and replated on fresh feeder layer. The first 'splitting' is often best performed as a 1:1 well replating. Use preconditioned (temperature and CO_2) medium. Cultures need to be refed once per day with new media (2–2.5 ml for a 6-well plate or 0.5 ml per 4-well plate). It is sometimes beneficial to carefully wash cells with hES medium to remove debris from the cultures.

Subsequent splittings can be made either mechanically or using mild enzyme treatment in combination with mild mechanical separation. Individual colonies are selected by the criterion of a uniform undifferentiated morphology. Early-stage human ES cell colonies

---

[2] (6): 9–15 days; (7): 6–8 days; (22): 7 days; (23): 9–19 days; (24): 7–10 days; (25): 8 days; (26): 4–8 days; (27): 4–11 days.
[3] Store diluted enzyme at +4, discard after 2–3 weeks.

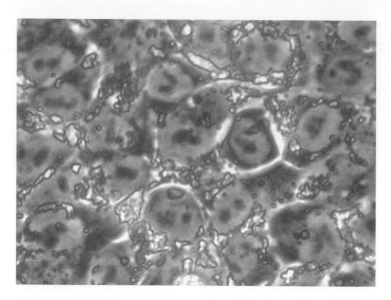

Figure 7.1

A human ES cell colony.

contain small round cells with prominent nucleoli as shown in *Figure 7.1*. Spontaneous differentiation is frequently observed during routine passage of the cells. Undesirable colonies can be separated with a pulled glass pipette.

The resulting colonies can be further propagated in clumps of ~100 cells, but usually not less. With time smaller colonies are possible to plate. At this early stage of culture it has not yet been shown possible to obtain a clonal expansion of a single cell. The initial colonies often have a less flat (somewhat three-dimensional) appearance which can fluctuate between passages. The doubling time of individual colonies can vary but the early colonies need to be passaged with a span of 3–14 days. Daily inspections of the cultures are vital and splitting can not be predecided.

For the cryopreservation of early cultures the method described by Reubinoff *et al.* (7) is particularly useful in that small numbers of cells in single colonies can be successfully cryopreserved.

NOTES

In vitro culture and differentiation of human embryonic stem cells

8

Maya Schuldiner and Nissim Benvenisty

8.1   Introduction

The study of mammalian cellular processes was revolutionized with the establishment of tissue culture techniques, allowing for long-term maintenance of mammalian cell lines. Unlimited growth of the cells was made possible thanks to the use of transformed cell lines and today can also be achieved by genetic manipulations, such as insertion of active telomerase into primary cells (1,2). One of the unique types of cell lines formed in the course of tissue culture history is embryonic stem (ES) cell lines (3), that are immortal though untransformed. The special characteristics of ES cells are a consequence of their tissue of origin. They are produced from the inner cell mass (ICM) of blastocyst-stage embryos and retain their embryonic qualities throughout prolonged culture. This means that undifferentiated ES cells can proliferate without limitation. These ES cell lines also have the capacity to differentiate into a wide variety of tissues from the three embryonic germ layers. These characteristics make them a very important tool in tissue culture research and as a model system to study early differentiation processes. The first mammalian ES cell lines to be established were from murine embryos (4,5). Though many ES-like cell lines are available today from other mammals (for a review see 6), none caused such a worldwide interest and excitement as the production of human ES cell (hES) lines (7,8). The reasons for this attention lie primarily in their biomedical potential as well as in their use as a unique model system for early human development and cellular differentiation. Such cell lines of human origin, maintaining their normal characteristics whilst still capable of infinite proliferation and the capacity to differentiate provide an unlimited cell source for transplantation medicine and toxicological testing (for reviews see 9,10).

In order for this potential to be utilized hES cells must be grown in a culture system that retains their pluripotent nature and their self-renewal potential. When a need for a differentiated cell type arises modified culture conditions are required that easily and consistently induce ES cells down that particular

Gene Targeting and Embryonic Stem Cells, Alison Thomson and Jim McWhir
© 2004 Garland Science/BIOS Scientific Publishers.

path. Many years of work with mouse ES cells have given a good basis of protocols for growing and differentiating ES cells (3). These protocols were modified to support the research with human ES cells.

8.2 Proliferation of human ES cells

The ICM of the blastocyst is a population of pluripotent cells, later differentiating into the three germ layers at the gastrulation stage. These transiently existing cells can be isolated and maintained in culture under tightly controlled culture conditions resulting in established lines of ES cells. Identifying the unique growth conditions required to maintain undifferentiated ES cells has been, and remains, central to the isolation and application of hES cells.

8.2.1 Effects of matrices on undifferentiated growth

The special growth conditions for maintaining undifferentiated proliferation rely mostly on the presence of a fibroblast feeder layer that was shown to support the pluripotent growth of ES cells. The feeder layer works at two levels, producing a growth matrix required to allow adherence of the cells, and secreting growth factors maintaining pluripotency and self-renewal potential (11). In mouse ES cells it was shown that both fresh mouse embryonic fibroblasts (MEFs) and immortalized cell lines (such as STO cells) can be used with similar efficiency (3). The necessity for the secreted component from the feeder layer for the growth of mouse ES cells was overcome with the identification of the effective cytokine they produced, namely leukemia inhibitory factor (LIF) (12,13). The addition of this cytokine allows maintenance of ES cell lines that are feeder-independent and still retain their pluripotent nature.

Human ES cells also rely on a fibroblast feeder layer for their undifferentiated growth. Unfortunately, despite the wide knowledge gained on the importance of LIF in maintenance of murine ES cells, initial evidence from human ES cell lines indicates that it does not play a critical role in sustaining undifferentiated growth (7,8). Additionally, some differences exist in the ability of fibroblast feeder layers to support self-renewal: immortalized cell lines such as STO do not give enough support for human ES cells (14) whereas primary cultures of MEFs do (8). Lately, it has been demonstrated that human fibroblasts either from embryos or from fallopian tubes are also suitable for sustaining growth and production of new cell lines (15). These findings necessitate the production of fresh MEFs for the culturing of human ES cells. Though optimal growth of human ES cells is dependent on a feeder layer, it has been shown that the chemical matrix 'matrigel' can be used as a substitute when combined with conditioned media from MEF cultures (14). Additionally, for a few passages, human ES cells can be maintained on much simpler chemical matrixes such as gelatin, collagen, fibronectin or poly-lysine. This is especially important when elimination of feeders is required before procedures such as differentiation into EBs, molecular and biochemical analysis or transplantation of cells derived from human ES cells. A comprehensive comparison of the effect of different matrices has not been performed, but it has been shown that laminin is superior to collagen IV or fibronectin in its ability to sustain pluripotent self renewal for many passages (14).

8.2.2 Significance of media components

In addition to the important role of the feeder cells in maintaining pluripotency, there is much influence of the basic growth media used for the propagation of human ES cells. The medium is very similar to that used for mouse ES cells. Medium and serum are a special formulation defined for the culturing of murine ES cells (16,17) though some laboratories use the standard Dulbeco's Modified Eagles Media with fetal bovine serum (7,8). Addition of β-mercaptoethanol seems to be an important constituent of media for stem cells (18), perhaps due to its antioxidant activity, that reduces the rate of redox-induced apoptosis (19). In addition to these, human ES cell growth depends on bFGF (8) though it is not clear whether the support is direct or through that of the MEF layer. In the case of cells grown on matrigel with conditioned medium the role of bFGF is still unclear as the MEFs used to condition the medium are also grown in bFGF.

8.2.3 Analysis of undifferentiated growth

The nature of ES cells is to differentiate, as they do in the context of the normal developing embryo. As described above, due to the specialized conditions employed in routine hES culture, the majority of the cells maintain undifferentiated growth. Nevertheless, a fraction of the population always escapes this fate and differentiates, especially at the periphery of the colony. These cells may cause further differentiation of the cultures if allowed to propagate without control. For this reason special care must be taken when growing human ES cells. It is advised not to passage cells for many generations before refreezing as this may allow propagation of semidifferentiated derivatives. In addition, when passaging cells, care must be taken not to let the cell numbers drop below a certain density, as this increases their tendency to differentiate, possibly due to lack of autocrine signaling (20).

The differentiation status of the cultures should be followed daily by observation through a phase contrast microscope. An undifferentiated colony has well-defined edges (see *Figure 8.1A*) and consists of small cells of similar size (see *Figure 8.1B*), all with a pronounced nucleus and clear cellular borders (see *Figure 8.1C*). As differentiation begins, the cells in the center or the edges may enlarge and lose their characteristic ES appearance. A more objective, yet time consuming, method for following the status of the cultures is by quantifying the expression of several molecular markers. The most well-characterized marker for pluripotency is the transcription factor Oct3/4 (OTF3) (21), whose expression is down-regulated upon differentiation of human ES cells (22, for a review see 23). Additional markers include high telomerase activity, pronounced alkaline phosphatase activity and staining for the cell surface markers SSEA-3, SSEA-4, TRA-1-60

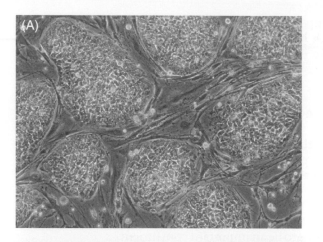

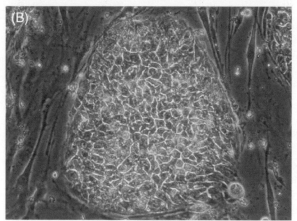

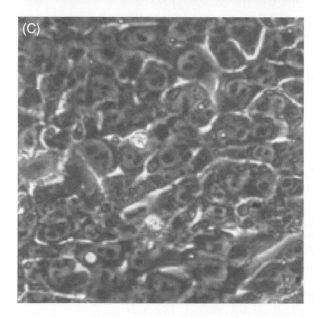

Figure 8.1

Human ES cell colonies. (**A**) Many colonies growing on a MEF feeder layer. Notice the sharp edges of the colonies. (**B**) Structure of a single colony. (**C**) Magnification of the cells in the colony. Notice the bright edges of each cell and the pronounced nucleus.

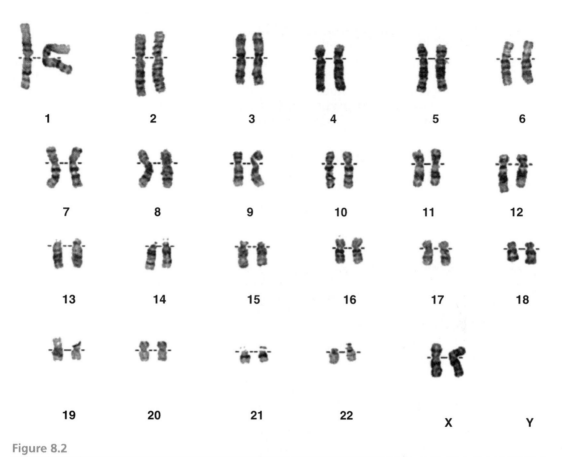

Figure 8.2

Normal 46XX karyotype of a human ES cell line.

and TRA-1-81 (8,14,24,25). One method to continuously monitor the undifferentiated status of human ES cells is by using cell lines expressing a marker gene under an ES-specific promoter. Such human ES cell lines have been established and express GFP under the *Rex1* promoter (26). If necessary, these cultures may be sorted using a fluorescent-activated cell sorter (FACS) to obtain pure populations of undifferentiated cells. Another possibility is to engineer cell lines expressing selection genes, such as neomycin resistance, under a stem cell-specific promoter thus making it possible to grow pure populations of ES cells when under constant selective pressure (27).

As with other cell lines growing *in vitro,* chromosomal aberrations may occur, taking over the population if they confer a growth advantage. This can be minimized by carefully controlling the number of passages that the cells grow in culture and by monitoring the karyotype of the cells following prolonged culture and after stable transfection (see *Figure 8.2*). As cultures containing a subpopulation of cells with an abnormal karyotype may occur in mouse and human ES cells (3,24), frozen stocks from early passages should always be maintained for use if growing cell cultures accumulate any chromosomal abnormality.

8.3 Differentiation of human ES cells

The pluripotent nature of ES cell lines is a unique characteristic, allowing them to differentiate into all cells from the three embryonic germ layers. Pluripotency is thus a characteristic trait of ES cells but is also a very important tool, allowing us to obtain large populations of differentiated cells for basic research or applicative uses.

8.3.1 Spontaneous differentiation

When grown under suboptimal conditions human ES cells tend to differentiate spontaneously. Under these conditions mouse ES cells have been shown to form mostly extraembryonic tissues (28). In order to consistently cause differentiation to all three embryonic germ layers it is preferable to transfer the human ES cells to plastic petri dishes to prevent adherence to the plate, thus allowing their aggregation into spheroid clumps termed embryoid bodies (EBs). The EBs develop through three stages: initially the cells form clusters called simple EBs (see *Figure 8.3A*). Later cells start dying at the center of the EBs resulting in the formation of cavitated EBs (see *Figure 8.3B*). Eventually, following 2 weeks of growth, they turn into cystic EBs with a fluid-filled cavity (see *Figure 8.3C*) (17). At this stage the EBs contain cells from various tissues. In some cases during formation of EBs from mouse ES cells it has been shown that development of cells from the hematopoietic system (29) and endothelia (30), molecularly recapitulate normal embryogenesis. This can be seen by the sequential activation of the globin or endothelial specific genes in the same order as was recorded in the developing embryo.

EBs can easily be formed by placing ES cells in a petri dish, but large variations in size and shape give rise to different outcomes using this method. In order to achieve a more homogenous EB culture with a known amount of cells, the 'hanging drop' method can be used (31). In the EBs many cell types spontaneously arise, though in an unorganized manner with no similarity to organ formation and with no detectable spatial organization. Work with human ES cells has shown that amongst other cell types, differentiating EBs contain muscle, bone, kidney, blood cells, skin, liver (16), β cells of the pancreas (16,32) and a large number of neurons (7,17,33,34). In some cases, the cells have even been shown to be functional as in the case of cardiomyocytes (17,35), neurons (34) or endothelial cells (36) (see *Figure 8.4*).

Spontaneous differentiation into various cell types may also occur without aggregation into EBs. This may be due to the spherical nature of overgrown ES colonies, that in some areas facilitate the three-dimensional interactions provided in the EB. In the neuronal lineage of the ectoderm this may be more pronounced due to the seemingly default nature of this differentiation pathway, requiring potentially less signals and cues to form (7,37).

8.3.2 Effects of growth factors

Human ES cells differentiate spontaneously into cells of the three germ layers and into a variety of different tissue types when allowed to aggregate and form EBs (16). However, this procedure is stochastic, giving rise to a mixture

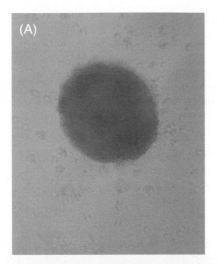

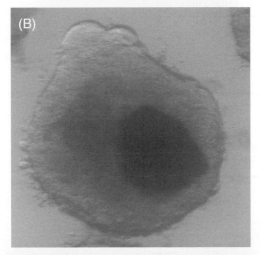

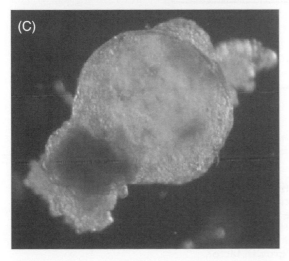

Figure 8.3

Human EBs in suspension. (**A**) Two-day-old simple EB. (**B**) Seven-day-old EB with a large dark cavitation at the lower right corner. (**C**) Fourteen-day-old, cystic, fluid filled EB.

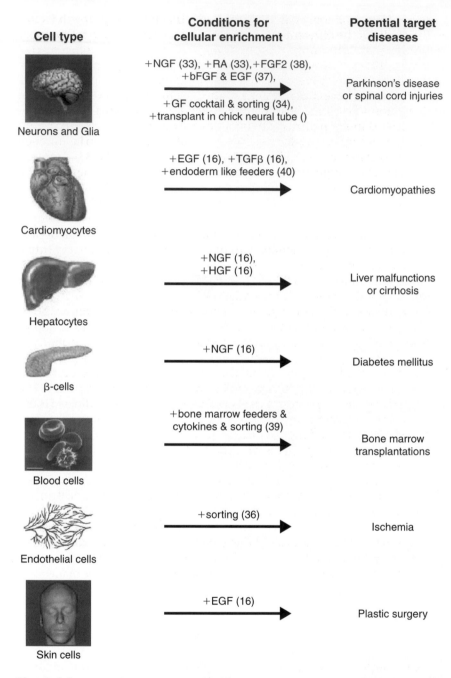

Figure 8.4

Induced differentiation of human ES cells. Schematic representation of methods for obtaining specific populations of enriched cells, and potential target diseases that may be treated by transplantation of each cell type. NGF: nerve growth factor; HGF: hepatocyte growth factor; TGFβ: transforming growth factor β; EGF: epidermal growth factor; RA: retinoic acid; bFGF: basic fibroblast growth factor.

of cells in various amounts. It has been shown that by adding growth factors to the media of the differentiating cells, they can be induced to follow more specific fates, thus forming a more homogenous population of differentiated cell types. This is a most important method for obtaining large numbers of differentiated cells for basic research or transplantation medicine. Neurons are the cells that have currently been studied the most intensively, partly due to their enormous biomedical potential to serve as a cell source in transplantation procedures of patients with neurological disorder such as Parkinson's disease. For example, it has been shown that mature neurons can be enriched by addition of nerve growth factor (NGF) or retinoic acid (RA) (33). Addition of a cocktail of many growth factors, including epidermal growth factor (EGF), basic fibroblast growth factor (bFGF), platelet-derived growth factor (PDGF), insulin-like growth factor (IGF1), neurotrophin 3 (NT3) and brain-derived neurotrophic factor (BDNF), allows more homogenous cultures to arise where functional neurons have been demonstrated (34). Enrichment of neurons, astrocytes and oligodendrocytes that are capable of integrating into a host brain, can also be achieved by addition of FGF2 (38) or by adding bFGF and EGF simultaneously (37) (see *Figure 8.4*). To evaluate the global effects of different growth factors on differentiation, cells derived from early human EBs were treated with various growth factors and markers for a dozen different cell types were monitored. Growth factors are large proteins, incapable of diffusing freely into a densely packed EB. One way to overcome this difficulty is to divide the differentiation process into two discrete steps, the first being aggregation into EBs for few days and the second, dissociation of the cells into a monolayer prior to administration of growth factors (see *Figure 8.5*) (16). Using this method, the factors each elicited a specific effect that divided them into three main groups: growth factors (activin-A and transforming growth factor β1 (TGFβ1)) that induce mainly muscular mesodermal cells; factors (RA, EGF, bone morphogenic protein-4 (BMP-4) and bFGF) that activate ectodermal and mesodermal markers; and factors (NGF and hepatocyte growth factor (HGF)) that allow differentiation into all three embryonic germ layers, including endoderm. Additionally each factor had some more explicit effects like enrichment of cultures with skin cells by EGF, liver cells by NGF or HGF and adrenal cells by RA (16) (see *Figure 8.4*). Until now none of these experiments showed clearly the basis of this enrichment. Thus, it could be due to induction of differentiation or to selection forces (negative or positive) of the growth factor acting on cells that have already been formed spontaneously in the cultures.

8.3.3   Influence of feeder layers and matrixes

Some cell types may be hard to obtain or to enrich solely by the addition of growth factors, as they require multiple signals in order to differentiate and/or survive. For this reason, cellular matrices of various kinds may be used. These feeder layers produce an assortment of diffusible growth factors in addition to cell surface signaling molecules. This has been tested for human ES cells in order to obtain diverse populations of hematopoietic cells (39) and cardiomyocytes (40) (see *Figure 8.4*). Yet, when using feeder layers, some thought has to be given to their elimination from the cultures at the end

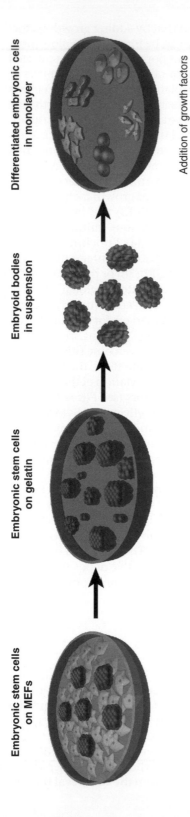

Figure 8.5

Schematic protocol for differentiation of human ES cells by addition of growth factors. Cells are grown on gelatin for a single passage to reduce contamination by mouse cells. Aggregation into simple EBs for approximately 4–5 days allows formation of the three embryonic germ layers. Dissociation and plating enables equal distribution of growth factors to the cell surface.

of the differentiation procedure. Chemical matrices also influence cellular signaling making it important to couple an optimal chemical matrix to the cells of choice. If no optimal growth factor combination or feeder layer can be found, it may be necessary to obtain specific cell types by transplanting the human ES cells into appropriate places in an animal model, allowing their differentiation according to local cues and later extracting them. Indeed, transplantation of human ES cells next to the chick neural tube has been shown to be a possible method of obtaining neuronal cells *in vivo* (41).

8.3.4   Cell sorting and selection

Though some protocols give rise to a large percentage of cells from a specific lineage, none has shown an ability to produce homogenous cultures of single cell types. This is probably due to the complexity of the signals both in time and space required for proper differentiation to occur *in vivo*. For this reason it is important to establish ways of sorting out cells of interest. This is especially crucial for transplantation procedures requiring a single cell type or for toxicological tests for a specific tissue, but is also an important step for the study of the intricate signaling mechanisms during specification of different cell fates and tissue maturation. Two main strategies are commonly used. The first requires genetic manipulation of the cells to form lines expressing a reporter or a selection gene under a tissue-specific promoter, and the second utilizes cell surface proteins to 'pull down' cells of choice. The first method was performed successfully with human ES cells by producing cell lines that express GFP under regulation of an ES cell specific promoter. The fluorescent cells could then be sorted out using a fluorescence-activated cell sorter (FACS), giving rise to a homogenous population of undifferentiated ES cells for many passages (see *Figure 8.6*) (26). A similar method can be used for any tissue of choice depending on the availability of promoter or enhancer sequences unique to a specific cell type or subset of cells as was shown for mouse ES-derived neuronal cells (42). Genetic manipulation also allows introduction of selection genes such as neomycin resistance under a cell-specific promoter, allowing clean cultures to be obtained without the need for cell sorting. This was previously shown for mouse ES cells using a cardiomyocyte specific promoter (43) or an insulin promoter to isolate β cells of the pancreas (44). The second strategy to isolate specific cell types utilizes antibodies against membrane proteins already present in the differentiating cells as targets for monoclonal antibodies. Such a procedure was demonstrated for purification of endothelial cells from human ES cells by the use of magnetic beads (36) and purification of human ES-derived hematopoietic cells was achieved by the use of a cell sorter (39). This procedure requires that the cell types of choice express a unique membrane marker and is therefore especially potent for cells of the hematopoietic system whose cell surface markers are well characterized.

8.4   Future prospects

Studies in the past few years of work with human ES cells have demonstrated their unique properties of unlimited proliferation and pluripotency. This

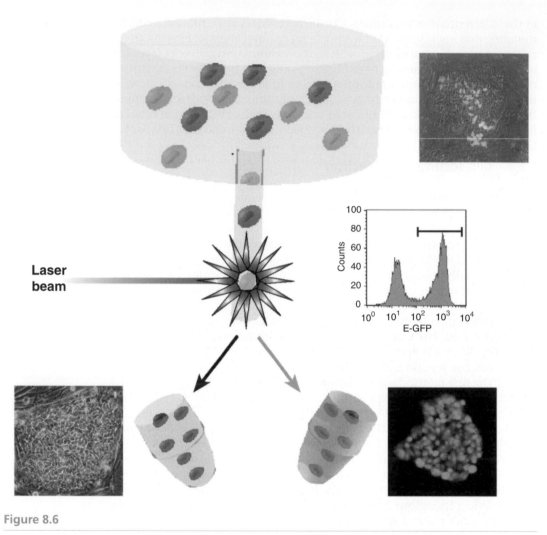

Figure 8.6

Sorting procedure of human ES cells using a FACS. Notice the primary population that consists of normal cells and GFP-positive cells. Based on fluorescence intensity, a threshold is determined and GFP-positive cells above that threshold are sorted out to give a clean population.

research also established methods for retaining the undifferentiated nature of human ES cells and controlling their differentiation. As one of the prospective uses of human ES cells includes transplantation, a major prerequisite is the identification of the specific factors that are involved in self-renewal that will allow the long-term culturing of human ES cells in the absence of a feeder layer. This will depend on characterization of the optimal chemical matrices required for their growth and the discovery of the important signaling molecules secreted by the MEF feeder layer. These discoveries will not only make clinical trials easier and safer due to lower contamination of feeder cells, but also will make routine culturing of the cells easier and more reproducible. Finally, the characterization of the signal transduction pathways important for self-renewal has profound implications for our understanding of the biological process of stem cell production and maintenance in the early embryo (for a review see 45).

Normal karyotype and lack of oncogenic mutations are important factors if medical uses are to be realized. As prolonged culture, even under optimal conditions, may give rise to chromosomal aberrations, or less easily detected point mutations, it may be important for clinical applications that only low passages of cells will be used. This will probably require routine production of additional cell lines for these procedures. At present, if cells are to be used for transplants they should be analyzed for a normal karyotype. Prospects for the future include the formation of DNA chips that will allow rapid profiling of gene expression in the transplanted cells.

The study of hES differentiation has relied mainly on prior knowledge from mouse ES cell investigations and research on other stem cell populations. In order for human ES cells to serve as a truly powerful tool for cellular transplantation, a much wider knowledge must be gained. This will include the refinement of techniques for obtaining enriched cultures of many cell types from different lineages, optimization of sorting techniques and the establishment of functionality and engraftment in a model system. Expanding the range of cell types we can isolate from hES cells may be achieved by analysis of the effects of additional growth factors under varying conditions and in combination with specialized matrices. An additional method for inducing differentiation is by manipulating the cells to express a specific transcription factor that plays a major role in cell commitment to a specific lineage, thus causing a large population to follow a specific fate path. This has previously been performed with mouse ES cells in order to induce differentiation to the otherwise elusive pathway of hepatic cells (46). Finally, it has been shown that differentiation induces up-regulation of the MHC molecules that govern tissue rejection (47). Thus, as future uses of human ES cells allow the formation of pure populations of differentiated cells, much thought will have to be given to the problem of tissue rejection (for a review see 9).

References

1. Bodnar AG, Ouellette M, Frolkis M, *et al.* (1998) Extension of life-span by introduction of telomerase into normal human cells. *Science* **279**: 349–352.
2. Yang J, Chang E, Cherry AM, *et al.* (1999) Human endothelial cell life extension by telomerase expression. *J Biol Chem* **274**: 26141–26148.
3. Robertson EJ (1987) Embryo-derived stem cell lines. In: Robertson EJ (ed) *Teratocarcinomas and Embryonic Stem Cells: A Practical Approach*, pp. 71–112. IRL Press, Washington DC.
4. Martin GR (1981) Isolation of a pluripotent cell line from early mouse embryos cultured in medium conditioned by teratocarcinoma stem cells. *Proc Natl Acad Sci USA* **78**: 7634–7638.
5. Evans MJ, Kaufman MH (1981) Establishment in culture of pluripotential cells from mouse embryos. *Nature* **292**: 154–156.
6. Prelle K, Vassiliev IM, Vassilieva SG, Wolf E, Wobus AM (1999) Establishment of pluripotent cell lines from vertebrate species – present status and future prospects. *Cell Tiss Org* **165**: 220–236.
7. Reubinoff BE, Pera MF, Fong CY, Trounson A, Bongso A (2000) Embryonic stem cell lines from human blastocysts: somatic differentiation in vitro. *Nat Biotechnol* **18**: 399–404.
8. Thomson JA, Itskovitz-Eldor J, Shapiro SS, Waknitz MA, Swiergiel JJ, Marshall VS, Jones JM (1998) Embryonic stem cell lines derived from human blastocysts. *Science* **282**: 1145–1147.
9. Schuldiner M, Benvenisty N (2001) Human embryonic stem cells: from human embryogenesis to cell therapy. In: Pandalai S (ed) *Recent Research and Developments in Molecular and Cellular Biology*, pp. 223–231. Research Signpost, Trivandrum.
10. Pera MF (2001) Human pluripotent stem cells: a progress report. *Curr Opin Genet Dev* **11**: 595–599.
11. Smith AG, Hooper ML (1987) Buffalo rat liver cells produce a diffusible activity which inhibits the differentiation of murine embryonal carcinoma and embryonic stem cells. *Dev Biol* **121**: 1–9.
12. Moreau JF, Donaldson DD, Bennett F, Witek-Giannotti J, Clark SC, Wong GG (1988) Leukaemia inhibitory factor is identical to the myeloid growth factor human interleukin for DA cells. *Nature* **336**: 690–692.
13. Smith AG, Heath JK, Donaldson DD, Wong GG, Moreau J, Stahl M, Rogers D (1988) Inhibition of pluripotential embryonic stem cell differentiation by purified polypeptides. *Nature* **336**: 688–690.
14. Xu C, Inokuma MS, Denham J, Golds K, Kundu P, Gold JD, Carpenter MK (2001) Feeder-free growth of undifferentiated human embryonic stem cells. *Nat Biotechnol* **19**: 971–974.
15. Richards M, Fong CY, Chan WK, Wong PC, Bongso A (2002) Human feeders support prolonged undifferentiated growth of human inner cell masses and embryonic stem cells. *Nat Biotechnol* **20**(9): 933–936.
16. Schuldiner M, Yanuka O, Itskovitz-Eldor J, Melton DA, Benvenisty N (2000) Effects of eight growth factors on the differentiation of cells derived from human embryonic stem cells. *Proc Natl Acad Sci USA* **97**: 11307–11312.
17. Itskovitz-Eldor J, Schuldiner M, Karsenti D, *et al.* (2000) Differentiation of human embryonic stem cells into embryoid bodies comprising the three embryonic germ layers. *Mol Med* **6**: 88–95.
18. Oshima R (1978) Stimulation of the clonal growth and differentiation of feeder layer dependent mouse embryonal carcinoma cells by beta-mercaptoethanol. *Differentiation* **11**: 149–155.
19. Coe JP, Rahman I, Sphyris N, Clarke AR, Harrison DJ (2002) Glutathione and p53 independently mediate responses against oxidative stress in ES cells. *Free Radic Biol Med* **32**: 187–196.

20. Dani C, Chambers I, Johnstone S, *et al.* (1998) Paracrine induction of stem cell renewal by LIF-deficient cells: a new ES cell regulatory pathway. *Dev Biol* **203**: 149–162.

21. Nichols J, Zevnik B, Anastassiadis K, *et al.* (1998) Formation of pluripotent stem cells in the mammalian embryo depends on the POU transcription factor Oct4. *Cell* **95**: 379–391.

22. Scholer HR, Dressler GR, Balling R, Rohdewohld H, Gruss P (1990) Oct-4: a germline-specific transcription factor mapping to the mouse t-complex. *EMBO J* **9**: 2185–2195.

23. Pesce M, Anastassiadis K, Scholer HR (1999) Oct-4: lessons of totipotency from embryonic stem cells. *Cell Tiss Org* **165**: 144–152.

24. Amit M, Carpenter MK, Inokuma MS, *et al.* (2000) Clonally derived human embryonic stem cell lines maintain pluripotency and proliferative potential for prolonged periods of culture. *Dev Biol* **227**: 271–278.

25. Draper JS, Pigott C, Thomson JA, Andrews PW (2002) Surface antigens of human embryonic stem cells: changes upon differentiation in culture. *J Anat* **200**: 249–258.

26. Eiges R, Schuldiner M, Drukker M, Yanuka O, Itskovitz-Eldor J, Benvenisty N (2001) Establishment of human embryonic stem cell-transfected clones carrying a marker for undifferentiated cells. *Curr Biol* **11**: 514–518.

27. Mountford P, Nichols J, Zevnik B, O'Brien C, Smith A (1998) Maintenance of pluripotential embryonic stem cells by stem cell selection. *Reprod Fertil Dev* **10**: 527–533.

28. Mummery CL, Feyen A, Freund E, Shen S (1990) Characteristics of embryonic stem cell differentiation: a comparison with two embryonal carcinoma cell lines. *Cell Differ Dev* **30**: 195–206.

29. Lindenbaum MH, Grosveld F (1990) An in vitro globin gene switching model based on differentiated embryonic stem cells. *Genes Dev* **4**: 2075–2085.

30. Vittet D, Prandini MH, Berthier R, Schweitzer A, Martin-Sisteron H, Uzan G, Dejana E (1996) Embryonic stem cells differentiate in vitro to endothelial cells through successive maturation steps. *Blood* **88**: 3424–3431.

31. Maltsev VA, Rohwedel J, Hescheler J, Wobus AM (1993) Embryonic stem cells differentiate in vitro into cardiomyocytes representing sinusnodal, atrial and ventricular cell types. *Mech Dev* **44**: 41–50.

32. Assady S, Maor G, Amit M, Itskovitz-Eldor J, Skorecki KL, Tzukerman M (2001) Insulin production by human embryonic stem cells. *Diabetes* **50**: 1691–1697.

33. Schuldiner M, Eiges R, Eden A, Yanuka O, Itskovitz-Eldor J, Goldstein RS, Benvenisty N (2001) Induced neuronal differentiation of human embryonic stem cells. *Brain Res* **913**: 201–205.

34. Carpenter MK, Inokuma MS, Denham J, Mujtaba T, Chiu CP, Rao MS (2001) Enrichment of neurons and neural precursors from human embryonic stem cells. *Exp Neurol* **172**: 383–397.

35. Kehat I, Kenyagin-Karsenti D, Snir M, *et al.* (2001) Human embryonic stem cells can differentiate into myocytes with structural and functional properties of cardiomyocytes. *J Clin Invest* **108**: 407–414.

36. Levenberg S, Golub JS, Amit M, Itskovitz-Eldor J, Langer R (2002) Endothelial cells derived from human embryonic stem cells. *Proc Natl Acad Sci USA* **99**: 4391–4396.

37. Reubinoff BE, Itsykson P, Turetsky T, Pera MF, Reinhartz E, Itzik A, Ben-Hur T (2001) Neural progenitors from human embryonic stem cells. *Nat Biotechnol* **19**: 1134–1140.

38. Zhang SC, Wernig M, Duncan ID, Brustle O, Thomson JA (2001) In vitro differentiation of transplantable neural precursors from human embryonic stem cells. *Nat Biotechnol* **19**: 1129–1133.

39. Kaufman DS, Hanson ET, Lewis RL, Auerbach R, Thomson JA (2001) Hematopoietic colony-forming cells derived from human embryonic stem cells. *Proc Natl Acad Sci USA* **98**: 10716–10721.
40. Mummery C, Ward D, van den Brink CE, *et al.* (2002) Cardiomyocyte differentiation of mouse and human embryonic stem cells. *J Anat* **200**: 233–242.
41. Goldstein RS, Drukker M, Reubinoff BE, Benvenisty N (2002) Integration and differentiation of human embryonic stem cells transplanted to the chick embryo. *Dev Dyn* **225**: 80–86.
42. Li M, Pevny L, Lovell-Badge R, Smith A (1998) Generation of purified neural precursors from embryonic stem cells by lineage selection. *Curr Biol* **8**: 971–974.
43. Klug MG, Soonpaa MH, Koh GY, Field LJ (1996) Genetically selected cardiomyocytes from differentiating embronic stem cells form stable intracardiac grafts. *J Clin Invest* **98**: 216–224.
44. Soria B, Roche E, Berna G, Leon-Quinto T, Reig JA, Martin F (2000) Insulin-secreting cells derived from embryonic stem cells normalize glycemia in streptozotocin-induced diabetic mice. *Diabetes* **49**: 157–162.
45. Burdon T, Chambers I, Stracey C, Niwa H, Smith A (1999) Signaling mechanisms regulating self-renewal and differentiation of pluripotent embryonic stem cells. *Cell Tiss Org* **165**: 131–143.
46. Levinson-Dushnik M, Benvenisty N (1997) Involvement of hepatocyte nuclear factor 3 in endoderm differentiation of embryonic stem cells. *Mol Cell Biol* **17**: 3817–3822.
47. Drukker M, Katz G, Urbach A, *et al.* (2002) Characterization of the expression of MHC proteins in human embryonic stem cells. *Proc Natl Acad Sci USA* **99**: 9864–9869.
48. Reubinoff BE, Pera MF, Vajta G, Trounson AO (2001) Effective cryopreservation of human embryonic stem cells by the open pulled straw vitrification method. *Hum Reprod* **16**: 2187–2194.

Protocols

Contents

MATERIALS

Reagents

Media reagents	KnockOut™ DMEM – Optimized Dulbecco's modified Eagle's medium for ES cells – store at 4°C
	DMEM 4.5 mg/ml glucose – store at 4°C
	KnockOut™ SR – serum-free formulation. Store aliquots at −20°C
	Fetal calf serum – store aliquots at −20°C
	200 mM glutamine stock – store aliquots at −20°C
	β-mercaptoethanol 1 M – store at 4°C
	100 × nonessential amino acids stock – store at 4°C
	Penicillin (10 000 U/ml) and streptomycin (10 mg/ml) stock – store aliquots at −20°C
	Basic fibroblast growth factor (bFGF) – store aliquots at −20°C
Additional reagents	Phosphate buffered saline (without Ca^{2+} or Mg^{2+}) – store at 4°C
	Bovine serum albumin – store at 4°C
	Gelatin powder – store at room temperature (RT)

Trypsin/EDTA 0.1%/1 mM – store at −20°C

DMSO – store at RT

Mitomycin C – store at 4°C

Equipment

Laminar flow hood

Humidified incubator set at 37°C with 5% CO_2

Swing out centrifuge for conal tubes

Phase contrast microscope

Tissue culture plates of various sizes

Tissue culture sterile pipettes of all sizes (1 ml, 2 ml, 5 ml, 10 ml, 25 ml) and pipettor

Sterile pasteur pipettes

Sterile pipette and diposable tips

Sterile Eppendorfs

Tissue culture cryotubes (1.5 ml) for freezing cells

Freezer set to −70°C

Liquid nitrogen storage racks immersed in liquid nitrogen

Isopropanol freezing device

Sharp operating equipment (tweezers, scissors)

Disposable syringes with 18G needles

Protocol 8.1: Preparation of media

PREPARATION OF bFGF STOCK

Dissolve every 10 μg bFGF in 5 ml filtered 0.1% BSA solution in PBS (2 ng/μl). Store individual aliquots (1 ml) at −20°C.

PREPARATION OF GELATIN STOCK

Add 0.1 g of gelatin into 100 ml H_2O (final conc. 0.1%) and autoclave immediately. The gelatin is dissolved whilst boiling in the autoclave. Store at 4°C.

GELATIN COATING OF PLATES

In laminar flow hood, to each 10 cm plate add 6 ml gelatin solution (or enough to cover plate surface). Let stand for a minimum of 1 hour. Aspirate remaining fluid until plate is dry. A thin film of gelatin will remain on the plate surface. Immediately apply appropriate media.

PREPARATION OF ES AND EB MEDIA

Prepare media in laminar flow hood by combining the following reagents. Store at 4°C for up to 1 month covered in foil (serum replacement is sensitive to light). Final concentrations are shown in parentheses.

400 ml KnockOut™ DMEM

80 ml KnockOut™ SR

5 ml (2 mM) glutamine

50 μl (0.1 mM) β-mercaptoethanol

5 ml (1×) nonessential amino acids stock

2.5 ml (50 units/ml) penicillin (50 μg/ml) streptomycin

For ES media only also add:

1 ml (4 ng/ml) basic fibroblast growth factor (bFGF)

PREPARATION OF MEF MEDIA

Prepare media in laminar flow hood by combining the following reagents. Store at 4°C for up to 1 month. Final concentrations are shown in parentheses.

450 ml DMEM

50 ml FCS

5 ml (2 mM) glutamine

2.5 ml (50 units/ml) penicillin (50 μg/ml) streptomycin

NOTES

Protocol 8.2: Passaging of human ES cells

The amounts are for 10 cm plates but can be up/down scaled accordingly. Work in laminar flow hood.

1. Aspirate medium from plate.

2. Wash with 5 ml PBS and aspirate.

3. Add 1 ml trypsin.

4. Tap plate on sides gently until cells are released. You may ascertain that cells are dissociated by looking under a phase contrast microscope.

5. Add 4 ml growth media and transfer to a conal tube.

6. Centrifuge 5 min at 2000 rpm.

7. Aspirate fluid and resuspend in growth medium.

8. Transfer to new plates at a dilution of 1:3 to 1:5.

Protocols for dissociation of cells using dispase or collagenase have also been published (8). Dissociation with these enzymes is much more delicate resulting in higher survival rates of the cells but leaving un-dissociated clumps. These clumps of cells proliferate less rapidly and have a tendency to differentiate. In addition, inability to dissociate cultures into single cells may prove problematic for transfection or sorting methods. Manual passaging of whole colonies and mechanical dissociation into smaller clumps is also possible (7) though labor intensive, and necessitates a specialized method for cryopreservation (48).

NOTES

Protocol 8.3: Freezing of MEFs and human ES cells

The quantities are for full 10 cm plates but can be up/down scaled accordingly. Work in laminar flow hood.

1. Follow passaging protocol up to stage 6.

2. Aspirate medium and resuspend in 900 μl media for each planned cryotube.

3. Add 100 μl DMSO for each 900 μl media, gently mix by pipetting up and down and transfer to properly marked cryotubes.

4. Transfer to − 70°C for 1 day. This step should preferably be done using a freezing container filled with isopropanol.

5. Transfer to liquid nitrogen for long-term storage.

NOTES

Protocol 8.4: Thawing of MEFs and human ES cells

1. Take out tube from liquid nitrogen and thaw as quickly as possible in water at 37°C. Work in laminar flow hood.

2. Transfer to a conal tube containing 4 ml of medium.

3. Centrifuge for 5 min in 2000 rpm.

4. Aspirate medium and resuspend in growth medium.

5. Transfer to tissue culture plates covered either with gelatin or mitomycin-C treated MEFs respectively.

NOTES

Protocol 8.5: Preparation of MEF cells

Use 13.5-day-old fetuses (13 days from morning of plug) from ICR mice that have not been hormone treated. Work on bench. Use sterile operating equipment.

1. Sacrifice a mouse and fix it on a surgery board with abdominal side up.

2. Sterilize the outer abdominal area using ethanol and let dry.

3. Open the stomach by ripping the skin or cutting both the skin and peritoneal cavity lining.

4. Take out the uterus from one end to the other with sterilized equipment. Most ICR females have between 10–14 fetuses in a single pregnancy.

5. Put the uterus in a clean petri dish.

6. Transfer all work to laminar flow hood.

7. Wash three times in 10 ml PBS or until clean of blood.

8. Using sterile tweezers and scissors separate the fetuses from the uterus, remove from amniotic sac and yolk sac and transfer to a clean petri dish – count the number of fetuses for later plating. (Prepare for later one gelatin-coated 10 cm tissue culture plate for each three clean fetuses.)

9. Wash three times in 10 ml PBS (be careful not to aspirate the fetuses).

10. Sterilize tweezers and scissors in ethanol and fire.

11. Remove the inner parts of each fetus by pulling gently at the red mass in the center of the embryo using two tweezers.

12. Remove the head of each fetus using scissors.

13. Transfer all cleaned fetuses to a new plate and add 2 ml of trypsin.

14. Using sharp scissors chop up the fetuses into small pieces.

15. Add 5 ml more of trypsin and put for 10 min in incubator (Instead of trypsinization it is also possible to pass the pieces ten times through a syringe with an 18G sterile needle. In this case suspend pieces in MEF media instead of trypsin and after passing through the syringe distribute them evenly into tissue culture plates. In both procedures big clumps are left, but many cells are dissociated.)

16. Add 10 ml medium and transfer into a 50 ml tube.

17. Wash plate with additional 5 ml medium and transfer to the tube.

18. Centrifuge for 5 min at 2000 rpm.

19. Remove upper liquid and resuspend in MEF media.

20. Distribute evenly between gelatin-coated 10 cm plates so that every plate gets some single cells and some clumps (the big clumps are important as they support the growth ability of the single cells).

21. Add medium to 10 ml in each plate and incubate.

22. Change media after 2 days.

23. When plates are confluent (2–3 days following production) passage each plate into three gelatin-coated plates. The passaging is important to eliminate the big clumps before freezing.

24. Change media every day, and when plates are confluent freeze each plate into a single cryotube.

25. When planning to treat with mitomycin-C, thaw contents of one cryotube onto three 10 cm plates. It is possible to passage the cells twice at a dilution of 1:3, so that 27 plates are obtained from each cryopreserved vial. At this stage all plates can be treated with mitomycin-C.

NOTES

The MEFs grow better on gelatin-coated plates so it is recommended to use coated plates when growing cells prior to mitomycin-C treatment. Following treatment it is important to use gelatin, as the MEFs will adhere better to gelatin-coated plates.

Protocol 8.6: Preparation of MEF feeder layers

MITOMYCIN-C TREATMENT OF MEFs

The protocol is aimed for MEFs grown on 10 cm plates. Work in laminar flow hood.

1. Aspirate medium and add 5 ml media containing 10 μg/ml mitomycin (50 μl from 1 mg/ml stock).

2. Incubate for 3 hours at 37°C.

3. Aspirate medium.

4. Wash twice with PBS to remove residual mitomycin-C.

5. Add 1 ml trypsin and put in incubator for 5 minutes.

6. Add 4 ml growth media and transfer to a conal tube.

7. Centrifuge for 5 min at 2000 rpm.

8. Aspirate top liquid and resuspend in growth medium.

9. Transfer 1–1.5 $\times$ 10^6 cells per 10 cm plates coated with gelatin. Cells are ready to use as a feeder layer the following day.

It is possible to freeze MEFs following treatment with mitomycin-C and keep stocks for later use. For this approximately 1.5–7.5 $\times$ 10^6 cells should be frozen in each cryotube and later thawed into 1–5 $\times$ 10 cm plates respectively.

NOTES

When using a feeder layer it is important to notice the coverage of the plate as it affects the growth of the ES cells. The perfect density is that in which there are the minimal number of cells that allow confluent coverage of the surface. Dense feeder layers compete with the ES cells and cause slow growth and differentiation.

Protocol 8.7: Formation of EBs

1. Passage human ES cells once on gelatin-coated plates in order to clean them from residual MEFs.

2. Harvest cells using trypsin, centrifuge and resuspend in EB medium (human ES cells media without bFGF).

For mass culture:

3. Place 10^7 cells into every 10 cm petri dish (it is recommended to UV irradiate plates before use).

4. Incubate for 2 days without moving plates.

5. Following 2 days, small aggregates are formed. At this stage change media in one of two ways:

   - Using a wide tip pipette in order to keep from damaging the EBs, gently transfer them to a conal tube, allowing them to settle *without* centrifugation. Aspirate media and resuspend in fresh media before transferring back to dish.

   - Place EB plate at an angle allowing EBs to concentrate at one end of the plate. After they have concentrated, aspirate as much of the media as possible and replace with the same amount of fresh media.

6. Media should be changed every 2–3 days (or more often if media becomes acidic) by the same method.

For 'hanging drop' method:

3. Count cells and suspend them in medium to a concentration of $1–10 \times 10^4$ cells/ml (400–4000 cells/40 μl).

4. On the cover of the tissue culture plate place 40 μl drops using a pipette (no more than 25 drops per lid).

5. Into the plate itself put 10 ml PBS in order to keep the drops from drying.

6. Place lid back onto plate so that the drops are hanging from the cover. (It is worthwhile practicing this step in advance with water drops.)

7. Place in incubator and do not touch or move for 2 days.

8. On the third day collect drops from lid using a pipette with a 1 ml tip cut at its end so as not to harm the EBs, and place in a petri dish with 10 ml medium.

9. Change medium every 2–3 days gently as described above.

ADMINISTRATION OF GROWTH FACTORS

1. Allow cells to form EBs as described above for 4 days.

2. Transfer into a conal tube, centrifuge, and remove upper liquid.

3. In order to dissociate EBs into single cells, resuspend in enough trypsin to cover completely and place in incubator for 10 minutes, shaking the tube every 2 minutes to assure dispersal of the clumps.

4. Add 5 ml of medium and centrifuge at 2000 rpm, remove upper liquid and resuspend in growth media.

5. Plate cells and remaining clumps on desired matrix (collagen, gelatin, fibronectin, polylysine, etc.).

6. Add growth factors to desired concentration.

7. Change media with fresh growth factors every 2 days.

NOTES

Transfection of human embryonic stem cells

9

Helen Priddle

9.1   Introduction

Since the first report of the isolation of human embryonic stem (ES) cells in 1998 (1) there has been a great deal of excitement about their potential. Confirmation of multipotency (2) and its maintenance after many population doublings (3) make human ES (hES) cells an ideal source of tissue for regenerative medicine. Additionally, hES cells provide a unique system with which to model human disease and development as well as for use in cytotoxicity testing. In order to realize this promise there are a number of potential problems to overcome. To bring nonautologous hES stem cell therapies to the clinic, rejection of hES-derived grafts by the patient's immune system must be avoided, programs of differentiation to provide therapeutically relevant cell types must be established and strategies to purify differentiated cell from a heterogeneous population, including potentially tumorigenic undifferentiated hES cells (1), must be developed. It is clear that the ability to genetically modify hES cells will be invaluable in the effort to surmount these hurdles.

9.1.1   The requirement for genetic modification of hES cells

The immune barrier to stem cell therapies could be circumvented in a number of ways (for review see 4) but the most practical solutions may be to render the hES cell and its derivatives 'invisible' to the immune system or render the immune system tolerant to the graft. The former may be possible by genetic modification of those loci of the hES cell genome responsible for immunogenicity. The latter involves tolerization of the patient to the hES cell line and its derivatives by the establishment of a chimeric immune system upon transplantation of hES cell-derived hematopoietic stem cells.

Strategies to optimize the production and purification of differentiated cell types, including hematopoietic stem cells, will be improved by a better understanding of human developmental biology. As hES cells are derived from the inner cell mass of the blastocyst, they provide a unique opportunity

Gene Targeting and Embryonic Stem Cells, Alison Thomson and Jim McWhir
© 2004 Garland Science/BIOS Scientific Publishers.

to study human development. Indeed, they have already been shown to give rise to trophoblast upon treatment with BMP4 (5). Dissection of pathways critical for commitment to specific lineages has been greatly enhanced by modification of genes of interest in mouse ES cells followed by *in vitro* differentiation to determine their role in development (e.g. see 6). A similar approach has also been employed to study the undifferentiated state in mouse ES cells (e.g. see 7). Such studies with human ES cells will not only be informative to human developmental biology but will enable cell biologists to manipulate gene pathways to improve differentiation to required cell types, as exemplified with mouse ES cells. Transgenes can be used to up-regulate pathways that promote commitment to the desired lineage (e.g. see 8) or to provide a lineage-specific marker to assist in the purification of a defined cell population (e.g. see 9).

Ensuring the destruction of potentially tumorigenic undifferentiated hES cells from a heterogeneous population of differentiated cells can be achieved using the expression of negatively selectable markers from an undifferentiated hES cell specific promoter. 'Suicide' gene therapy has already been developed for the treatment of cancer (10). Adaptation of this approach for hES cells will be less complex as the cells would be modified *in vitro* to express the suicide gene and do not require targeted delivery *in vivo*. Recently, some progress has been made towards this aim. Human ES cells were modified to constitutively express the thymidine kinase gene such that application of the prodrug gancyclovir caused ablation of hES cells *in vitro* and their derivatives *in vivo* in SCID mice (11).

The study of human disease can be improved by the introduction of disease-linked mutations into hES cells followed by *in vitro* differentiation and subsequent analysis of the affected cell type. The majority of disease models studied using transgenic mouse ES cells have involved the production of transgenic mice (e.g. see 12) an approach neither possible nor desirable in humans. However *in vitro* mouse disease models have been employed to explore both the phenotype of genetic diseases (e.g. see 13) and potential therapies (e.g. see 14). Moreover, it may be possible to alleviate genetic disease by targeted correction of disease genes. Controversially, therapeutic cloning could provide blastocysts genetically identical to the patient from which hES cells could be derived. Genetic modification of the disease allele would allow the corrected hES cells to be differentiated to a normal version of the affected cell type for transplant. Proof of principle has been achieved in an experimental system where a healthy immune system was restored to immune-deficient Rag2 mutant mice (15). However, in the face of the ethical and practical problems associated with therapeutic cloning and subsequent gene correction in humans, it seems likely that a transplant derived from normal, but allogeneic, hES cells will be less problematic.

9.1.2 The use of gene targeting in hES cells

Many of the genetic modifications described, such as the modeling of human disease, reduction of immunogenicity of hES cell-derived transplants and dissection of pathways in developmental biology, would specifically require the modification of hES cells by gene targeting. However, gene targeting is not necessary for the simple expression of an exogenous transgene, for example to increase commitment to a specific lineage. Random integration of coding sequence and regulatory elements into the hES cell genome would be sufficient to provide stable transgene expression. Arguably, random integration of transgenes can be dangerous due to the risk of insertional inactivation of coding sequences or perturbation of regulatory regions. In addition, random integration of transgenes can lead to reduced and erroneous expression due to position effects (16). Both of these problems can be overcome by the use of homologous recombination to introduce the transgene to a site known to be permissive for transgene expression (17). Thus gene targeting provides a safer and more effective route to stable transgene expression.

9.1.3   Problems associated with gene targeting in hES cells

The process of gene targeting in mouse ES cells typically involves the construction of a gene-targeting vector (including genomic DNA identical to the target locus and a selectable marker), transfection of the targeting vector into cells, antibiotic selection for transfected clones and screening a large number of clones to identify some which have integrated the targeting vector by homologous recombination (e.g. see 18). Homologous recombination in mouse ES cells is generally a rare event, and the frequency of gene-targeted clones (as a proportion of stably transfected clones) varies between different target loci, from as little as 1 in 20 000 (19) to exceeding 10% (e.g. see 20). Thus when early reports (21) suggested that standard transfection procedures with hES cells gave poor transfection efficiencies it became apparent this would be a major obstacle to the routine isolation of gene-targeted human ES cells. To date only the hypoxanthine phosphoribosyl transferase (HPRT) and Oct4 genes have been successfully targeted in hES cells (22). Transfection regimes may require further refinement for other gene-targeting applications.

   Different cell culture regimes have been employed by laboratories developing transfection procedures for hES cells (see *Table 9.1*). These range from manual disaggregation of hES cells (23) to trypsin-based disaggregation (21); from feeder-dependent (21) to feeder-free cultivation (25). Problems generating single-cell suspensions and poor plating efficiencies after passage have hampered progress, as has poor cloning efficiency. Additionally, the presence of feeders can give 'background' transfection and antibiotic selection of transfected clones is only possible if the feeders are antibiotic resistant (21).

9.2   Progress towards efficient transfection of hES cells

9.2.1   Improvements in cell culture regimes

The establishment of a feeder-free culture system (25) has been a major advance in hES technology, not least for hES transgenics. The requirement for growth on mouse embryonic fibroblast (MEF) feeder layers was replaced by growth on a complex extracellular matrix (Matrigel) in the presence of medium conditioned by MEF feeder cells. Thus it is now possible to harvest feeder-free hES cells for transfection and to apply antibiotic selection without the need for antibiotic-resistant feeders. Amit *et al.* (3) demonstrated that the cloning efficiency of hES cells could be improved if hES cells were plated at clonal density in serum-free medium in the presence of basic fibroblast growth factor (bFGF), thus improving the isolation of transgenic hES lines subsequent to transfection, low-density plating and antibiotic selection. Additionally, more researchers are using trypsin-based disaggregation of hES cells (see *Table 9.1*), which allows the preparation of a single-cell suspension which is more effective for electroporation. Other methods that involved scraping or other forms of mechanical disruption of cells can result in a low plating efficiency, which in turn diminishes the recovery of transfected clones.

9.2.2 Viral transduction

Exogenous DNA has been introduced into hES cells using a number of methods (see *Table 9.1*). These include viral, chemical (using lipids or cationic polymers) and physical (electroporation) methods. Transduction rates with lentiviral vectors have been impressive (see *Table 9.1*) and although efficiency seems lower for adenoviral and adeno-associated vectors, these have proved effective in gene-targeting experiments in human somatic cells (26). Reports of viral transduction have concentrated on introducing reporter genes (EGFP and LacZ) to cells and assessing the extent of transfection of a population (23,24,27,28). Whilst this approach can be valuable for hES cell lines where the plating of cells at clonal density is problematic, viral vectors have maxima with respect to the length of exogenous DNA that can be introduced, limiting their use in the construction of targeting vectors for complex genomic alterations. Accordingly, this chapter concentrates on chemical and physical transfection techniques.

9.2.3 Optimizing transfection

A number of independent groups have compared different chemical transfection reagents and electroporation techniques (see *Figure 9.1* and *Table 9.1*). Eiges *et al.* (21) assessed a number of techniques by comparing the subsequent level of luminosity resulting from expression of a transfected firefly Rennila transgene. They found the cationic polymer, ExGen 500 (Fermentas), to be most effective. This was used to produce the first reported transfected hES clones (expressing EGFP) with a transfection efficiency of 3.3×10^{-5} (21), although the efficiency was reduced in a later experiment by the same group (11). Zwaka and Thomson (22) tested Fugene (Roche) and ExGen 500 to give transfection efficiencies of approximately 1×10^{-5} and we have used Lipofectamine 2000 (Invitrogen) at an efficiency of 4.1×10^{-5} (see *Table 9.1* and *Figure 9.1*). More effective promoters can increase the recovery of G418-resistant colonies upon chemical transfection of a neo transgene. We have found that under these conditions

Table 9.1 Summary of transfections with human ES cells

Ref	Line	Culture details	Construct	Method	Parameters	Cells	Clones	T.E.
19	mES	Feeder free (LIF), trypsin	MC1-Neo	Electroporated	750 V/cm	1×10^8	5×10^4	5×10^{-3}
28	H9	Feeder free, trypsin/EGTA	RSV/LTR-LacZ	Adenoviral	500 MOI	nd	nd	11.2%
24	H9	Feeder free, trypsin/EDTA	EF1α-EGFP	Lentiviral	12 MOI	nd	nd	87%
27	H1	Feeders, collagenase	CAG-EGFP	Lentiviral	50 MOI	nd	nd	~100%
23	HES1	Feeders*, manual + dispase	EF1α-EGFP	Lentiviral	nd	nd	nd	30–48%
a	H9	Feeder free, trypsin/EGTA	PGK-Neo	Chemical	LF2000	1×10^6	41	4.1×10^{-5}
				Gene Pulser	200V/950 μFD	1×10^6	36	3.6×10^{-5}
				Multiporator	300V/100 μS	1×10^6	231	2.3×10^{-4}
				Multiporator	300V/100 μS	1×10^6	52	5.2×10^{-5}
b	H1	Feeder free, trypsin/EGTA	PGK-Neo	Chemical	Fugene	nd	261	$\sim 1 \times 10^{-5}$
			tk-Neo†		ExGen	nd	130	$\sim 1 \times 10^{-5}$
22	H1.1	Feeders*, collagenase	tk-Neo†	Gene Pulser	220V/960 μFD	nd	nd	$\sim 1 \times 10^{-7}$
					320V/200 μFD	1.5×10^7	350	2.3×10^{-5}
			EGFP/Neo†‡	Gene Puser	320V/200 μFD	1.5×10^7	103	6.8×10^{-6}
21	H9	Neo⁺ feeders, trypsin	tk-firefly renilla	Electroporated	nd	Relative rennila luminocity →		~20
					LF			~control
					Fugene			~20
					ExGen 500			~160
11	H9	Neo⁺ feeders, trypsin	SV40 Neo†	Chemical	ExGen 500	3×10^5	10	3.3×10^{-5}
			PGK-neo†	Chemical	ExGen 500	1×10^7	9	9×10^{-7}

a: see Figure 9.1; b: see Figure 9.2; T.E.: transfection efficiency; MOI: multiplicities of infection; nd: not determined or not stated; LF: Lipofectamine; *feeders were removed for transfection; †in the context of a more complex vector; ‡Coupled by IRES to Oct4 promoter.

(A)

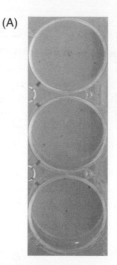

(B)

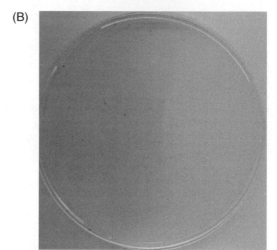

(C)

Figure 9.1

Comparison of different transfection methods.
A neomycin resistance gene under the
control of a PGK promoter was introduced
into 1×10^6 H9 hES cells by (**A**) lipofection,
(**B**) electroporation using the Gene Pulser
system with 200 V/950 μFD or (**C**) the
Multiporator system with 300 V/100 μS.
G418 selection (150 μg/ml) was applied 48
hours post-transfection for 2 weeks. G418
resistant colonies were fixed with methanol
and stained with 10% Giemsa.

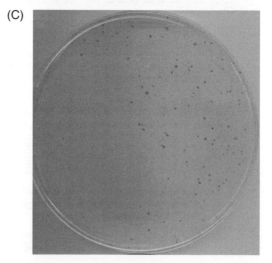

the Ef1α promoter and first intron (from pEF1/V5HisA, Invitrogen) gives the most colonies, followed by the hTERT promoter, the SV40 and PGK promoter (Priddle H, Thomson A, DiDomenico A, Wojtacha D, unpublished results). Whilst chemical transfection efficiencies are sufficient to generate transgenic hES cells lines, they fall a long way short of typical transfection efficiencies in mouse ES cells (e.g. 5×10^{-3} (19)) and would not provide sufficient clone numbers to optimize the chance of identifying gene-targeted clones from a transfection. Additionally, lipofection has been shown to be less effective than electroporation for the production of gene-targeted human somatic cells (29). Hence the focus of improvements for gene targeting in hES cells has been on electroporation methods.

Zwaka and Thomson (22) optimized electroporation for H1.1 hES cells cultured with feeders and passaged using collagenase (see *Table 9.1*). To avoid problems associated with feeder cells, they cultured on matrigel in the presence of feeder-conditioned medium from 1 week prior to transfection and throughout selection. As collagenase gives rise to clumps of ES cells, and as hES cells are larger than mouse ES cells, they used electroporation conditions designed for larger cells with the BioRad Gene Pulser II electroporator. Rather than electroporating their cells in low-protein buffer they electroporated in culture medium. By this method they succeeded in increasing the transfection efficiencies from 1×10^{-7} to around 2×10^{-5}. We took an alternative approach, using the Multiporator system (Eppendorf), which utilizes a hypo-osmolar buffer to expand the membrane and loosen its association with the cytoskeleton, reducing the required voltage. The Multiporator also applies the voltage in a shorter pulse, in the micro-second rather than millisecond range which is much gentler for the cells (see manufacturer's Basic Applications Guide). We optimized this system for H9 hES cells, in conjunction with feeder-free culture and trypsin-based disaggregation to a single cell suspension. We improved upon our best results with the Gene Pulser II electroporator (3.6×10^{-5}) by more than ten-fold (2.3×10^{-4}) in a controlled comparison (see *Figure 9.1* and *Table 9.1*). However, it is clear that the same conditions applied to the H1 hES cell line gave a lower transfection efficiency (5.2×10^{-5}; see *Figure 9.2* and *Table 9.1*). This may be explained by the lower plating efficiency of H1 hES cells under the conditions used, or a different cell diameter requiring a different field strength for membrane poration.

9.3   Future of gene targeting in human ES cells

Recent improvements in hES cell culture techniques and electroporation procedures have greatly enhanced the efficiency of stable transfection, leading to the recent report by Zwaka and Thomson (22) of successful gene targeting at two expressed loci in human ES cells, at frequencies similar to those seen at equivalent loci in mouse ES cells. Thus the future of gene targeting in human ES cells seems bright. However, there are still a number of issues to be resolved before it becomes clear how useful and routine this technique will be. Given the requirement for plating at clonal density and expansion through many population doublings, are these gene-targeted hES cell lines karyotypically stable and do they retain their multipotency? This was not discussed by Zwaka and Thomson (22), although some evidence from

(A)

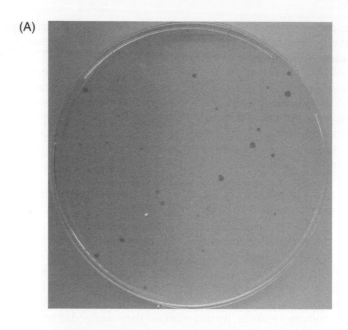

(B)

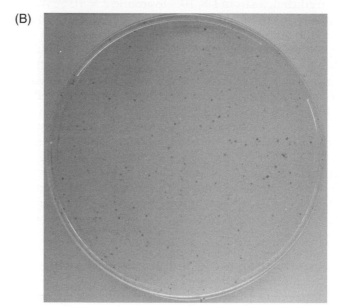

Figure 9.2

Comparison of transfection of different hES cell lines. A neomycin resistance gene under the control of a PGK promoter was introduced into (**A**) 1×10^6 H1 hES cells or (**B**) 1×10^6 H9 hES cells by electroporation using the Multiporator system with 300 V/100 μS. G418 selection (150 μg/ml) was applied 48 hours post-transfection for approximately 2 weeks. G418 resistant colonies were fixed with methanol and stained with 10% Giemsa.

nontargeted, transgenic hES lines suggests that such manipulations are not detrimental (11,21,24). Will gene targeting be equally effective for different hES cell lines? And if the technique is to become routine and applicable for loci subject to lower recombination frequencies, then we must continue to improve upon transfection efficiencies.

References

1. Thomson JA, Itskovitz-Eldor J, Shapiro SS, Waknitz MA, Swiergiel JJ, Marshall VS, Jones JM (1998) Embryonic stem cell lines derived from human blastocysts. *Science* **282**: 1145–1147.

2. Itskovitz-Eldor J, Schuldiner M, Karsenti D, *et al.* (2000) Differentiation of human embryonic stem cells into embryoid bodies compromising the three embryonic germ layers. *Mol Med* **6**: 88–95.

3. Amit M, Carpenter MK, Inokuma MS, *et al.* (2000) Clonally derived human embryonic stem cell lines maintain pluripotency and proliferative potential for prolonged periods of culture. *Dev Biol* **227**: 271–278.

4. Bradley JA, Bolton EM, Pedersen RA (2002) Stem cell medicine encounters the immune system. *Nat Rev Immunol* **2**: 859–871.

5. Xu RH, Chen X, Li DS, *et al.* (2002) BMP4 initiates human embryonic stem cell differentiation to trophoblast. *Nat Biotechnol* **20**: 1261–1264.

6. Hidaka M, Stanford WL, Bernstein A (1999) Conditional requirement for the Flk-1 receptor in the *in vitro* generation of early haematopoietic cells. *Proc Natl Acad Sci USA* **96**: 7370–7375.

7. Niwa H, Miyazaki J, Smith AG (2000) Quantitative expression of Oct-3/4 defines differentiation, dedifferentiation or self-renewal of ES cells. *Nat Genet* **24**: 372–376.

8. Kyba M, Perlingeiro RC, Daley GQ (2002) HoxB4 confers definitive lymphoidmyeloid engraftment potential on embryonic stem cell and yolk sac haematopoietic progenitors. *Cell* **109**: 29–37.

9. Ying QL, Stavridis M, Griffiths D, Li M, Smith A (2003) Conversion of embryonic stem cells into neuroectodermal precursors in adherent monoculture. *Nat Biotechnol* **21**: 183–186.

10. Freytag SO, Khil M, Stricker H, *et al.* (2002) Phase I study of replication-competent adenovirus-mediated double suicide gene therapy for the treatment of locally recurrent prostate cancer. *Cancer Res* **62**: 4968–4976.

11. Schuldiner M, Itskovitz-Eldor J, Benvenisty N (2003) Selective ablation of human embryonic stem cells expressing a "Suicide" gene. *Stem Cells* **21**: 257–265.

12. Davidson DJ, Dorin JR, McLachlan G, *et al.* (1995) Lung disease in the cystic fibrosis mouse exposed to bacterial pathogens. *Nat Genet* **9**: 351–357.

13. Metzler M, Chen N, Helgason CD, *et al.* (1999) Life without huntingtin: normal differentiation into functional neurons. *J Neurochem* **72**: 1009–1018.

14. Dunn DE, Yu J, Nagarajan S, *et al.* (1996) A knock-out model of paroxysmal nocturnal hemoglobinuria: Pig-a(-) haematopoiesis is reconstituted following intercellular transfer of GPI-anchored proteins. *Proc Natl Acad Sci USA* **93**: 7938–7943.

15. Rideout WM 3rd, Hochedlinger K, Kyba M, Daley GQ, Jaenisch R (2002) Correction of a genetic defect by nuclear transplantation and combined cell and gene therapy. *Cell* **109**: 17–27.

16. Clark AJ, Bissinger P, Bullock DW, Damak S, Wallace R, Whitelaw CB, Yull F (1994) Chromosomal position effects and the modulation of transgene expression. *Reprod Fertil Dev* **6**: 589–598.

17. Wallace H, Ansell R, Clark J, McWhir J (2000) Pre-selection of integration sites imparts repeatable transgene expression. *Nucleic Acids Res* **28**: 1455–1464.

18. Priddle H, Hemmings L, Monkley S, *et al.* (1998) Disruption of the talin gene compromises focal adhesion assembly in undifferentiated but not differentiated embryonic stem cells. *J Cell Biol* **142**: 1121–1133.

19. Johnson RS, Sheng M, Greenberg ME, Kolodner RD, Papaioannou VE, Spiegelman BM (1989) Targeting of nonexpressed genes in embryonic stem cells via homologous recombination. *Science* **245**: 1234–1236.

20. Domínguez-Bendala J, Priddle H, Clarke A, McWhir J (2003) Elevated expression of exogenous Rad51 leads to identical increases in gene-targeting frequency in murine embryonic stem (ES) cells with both functional and dysfunctional p53 genes. *Exp Cell Res* **286**: 298–307.

21. Eiges R, Schuldiner M, Drukker M, Yanuka O, Itskovitz-Eldor J, Benvenisty N (2001) Establishment of human embryonic stem cell-transfected clones carrying a marker for undifferentiated cells. *Curr Biol* **11**: 514–518.

22. Zwaka TP, Thomson JA (2003) Homologous recombination in human embryonic stem cells. *Nat Biotechnol* **21**: 319–321.

23. Gropp M, Itsykson P, Singer O, Ben-Hur T, Reinhartz E, Galun E, Reubinoff BE (2003) Stable genetic modification of human embryonic stem cells by lentiviral vectors. *Mol Ther* **7**: 281–287.

24. Ma Y, Ramezani A, Lewis R, Hawley RG, Thomson JA (2003) High-level sustained transgene expression in human embryonic stem cells using lentiviral vectors. *Stem Cells* **21**: 111–117.

25. Xu C, Inokuma MS, Denham J, Golds K, Kundu P, Gold JD, Carpenter MK (2001) Feeder-free growth of undifferentiated human embryonic stem cells. *Nat Biotechnol* **19**: 971–974.

26. Hirata R, Chamberlain J, Dong R, Russell DW (2002) Targeted transgene insertion into human chromosomes by adeno-associated virus vectors. *Nat Biotechnol* **20**: 735–738.

27. Pfeifer A, Ikawa M, Dayn Y, Verma IM (2002) Transgenesis by lentiviral vectors: lack of gene silencing in mammalian embryonic stem cells and preimplantation embryos. *Proc Natl Acad Sci USA* **99**: 2140–2145.

28. Smith-Arica JR, Thomson AJ, Ansell R, Chiorini J, Davidson B, McWhir J (2003) Infection efficiency of human and mouse embryonic stem cells using adenoviral and adeno-associated viral vectors. *Cloning Stem Cells* **5**: 51–62.

29. Yanez RJ, Porter AC (1999) Influence of DNA delivery method on gene targeting frequencies in human cells. *Somat Cell Mol Genet* **25**: 27–31.

Protocols

Contents

MATERIALS

All reagents are obtained from Sigma-Aldrich Co. Ltd., UK, except where otherwise stated.

PBS

Mg^{2+} and Ca^{2+} free. Make with four Dulbecco 'A' PBS tablets (Oxoid) in 400 ml double distilled water. Autoclave and store at room temperature.

HBS

0.476 g	HEPES
0.801 g	Sodium chloride
0.037 g	Potassium chloride
0.010 g	Disodium hydrogen phosphate
0.108 g	Glucose

Adjust pH to 7.05 and make up to 100 ml with double distilled water. Autoclave and store at room temperature.

TEG disaggregation reagent

6.30 g	Sodium chloride
0.12 g	Disodium hydrogen phosphate
0.22 g	Potassium dihydrogen phosphate

0.33 g	Potassium chloride
0.90 g	Glucose
2.70 g	Tris
0.90 ml	1% Phenol Red

Make up to 800 ml with double distilled water and add:

100 ml	2.5% (10×) Trypsin solution in saline (Invitrogen)
0.40 g	EGTA
0.10 g	Polyvinyl alcohol

Adjust pH to 7.6 and make up to 1000 ml with double distilled water. Filter sterilize, aliquot and store at –20°C.

Mouse embryonic fibroblast (MEF) medium

450 ml	DMEM with 4.5 g glucose per liter (Invitrogen)
50 ml	Fetal calf serum (FCS) (Globepharm)
5 ml	200 mM L-glutamine (Invitrogen)

Filter sterilize and store at 4°C for up to 4 weeks.

Serum-free hES medium

400 ml	Knockout (KO) DMEM (Invitrogen)
100 ml	KO serum replacement (Invitrogen)
5 ml	200 mM L-glutamine (Invitrogen)
5 ml	100 × nonessential amino acids (Invitrogen)
1 ml	50 mM β-mercaptoethanol (Invitrogen)

Filter sterilize and store at 4°C for up to 4 weeks.

bFGF stock solution (1 μg ml^{-1})

Dissolve 25 μg recombinant human bFGF (Sigma) in 25 ml PBS supplemented with 0.1% BSA. Filter sterilize through a 0.2 μM syringe filter. Make 1 ml aliquots and store at –20°C.

Matrigel

A 10 ml bottle of growth factor reduced Matrigel (Fahrenheit) should be thawed on ice at 4°C overnight, ensuring sufficient ice to last. Make 0.5 ml aliquots in prechilled tubes with a prechilled pipette. Do not allow the Matrigel to warm above 0°C as this triggers polymerization. Store at –20°C.

Methods

Protocol 9.1: Feeder-free culture of hES cells

This protocol is based on the original report of feeder-free hES cell culture (25).

PREPARATION OF MEFs

1. Sacrifice a pregnant balb/c mouse at day 13.5 of pregnancy, by cervical dislocation.

2. Dissect out uterine horns, dip them briefly in 70% ethanol and transfer into PBS (containing penicillin/streptomycin solution; Invitrogen) in a bacterial Petri dish.

3. Dissect out fetuses and decapitate. Remove red tissues from gut region.

4. Wash embryos through three fresh Petri dishes of PBS/pen/strep and transfer to a tube for transport to a tissue culture hood.

5. Incubate each embryo in 2 ml TEG solution, vortex, and incubate at 37°C for 5 minutes.

6. Repeat the vortex and incubation step sufficient times for the production of a dense 'soup-like' cell suspension.

7. Add 3 ml MEF medium, vortex and let larger debris settle for 1 minute.

8. Remove the top 3.5 ml of cell suspension and place in a 75 cm^2 tissue culture flask with a further 15 ml MEF medium (one flask for each embryo).

9. Incubate until confluent in a humidified 37°C incubator with 5% CO_2.

10. As each flask approaches confluence, wash with PBS, aspirate and incubate with 3 ml TEG until the MEFs can be knocked off the tissue culture surface.

11. Transfer cells into a 150 cm^2 tissue culture flask and add 40 ml MEF medium.

12. Incubate until confluent, wash with PBS and treat with 5 ml TEG as before.

13. Pool all cells and centrifuge at 200 *g* for 5 minutes. Resuspend in 1.5 ml MEF medium for every flask used.

14. Add an equal volume of chilled freezing mix (four parts MEF medium and one part DMSO), mix gently and place 1 ml aliquots in cryovials.

15. Place at −80°C overnight and then store under liquid nitrogen.

CONDITIONING hES MEDIUM

1. Rapidly thaw a cryovial of MEFs by placing in a 37°C water bath.

2. Immediately transfer the cells into 10 ml of prewarmed MEF medium and centrifuge at 200 *g* for 5 minutes.

3. Resuspend the MEFs in 20 ml MEF medium, transfer to a 75 cm^2 tissue culture flask and incubate until confluent.

4. MEFs can be passaged or expanded with TEG as described above, and are used up to passage 4 for conditioning of medium.

5. Use a 25 cm^2 flask of MEFs to condition 8 ml medium

    75 cm^2 flask of MEFs to condition 15 ml medium

    150 cm^2 flask of MEFs to condition 40 ml medium

6. Expand MEFs in MEF medium in an appropriately sized flask to approximately 80% confluence. Aspirate medium and replace with hES medium supplemented with a 1 in 250 dilution of bFGF stock per ml of hES medium (volumes above) to give a final concentration of 4 ng ml^{-1}.

7. Leave the flask 18–24 hours overnight for conditioning to take place, then harvest the medium and replace with fresh hES medium/bFGF for harvest the following day.

8. Add more bFGF stock to the harvested medium to 4 ng ml^{-1} and filter sterilize. This medium can be used directly for the culture of hES cells.

9. A monolayer of MEFs can be used up to four times.

10. Frozen conditioned medium can be used; we recommend storage at −20°C prior to the second addition of bFGF, which should be added after thawing to avoid its repeated freezing and thawing.

CULTIVATING hES CELLS ON MATRIGEL MATRIX IN CONDITIONED MEDIUM

1. Thaw an aliquot of Matrigel overnight on ice or at 4°C for 30 minutes.

2. Add to 50 ml ice cold KO DMEM with a prechilled pipette and mix immediately by inverting.

3. Use 0.2 ml per cm^2 over the tissue culture surface being prepared for hES cells and allow the matrigel to polymerize on a level surface at 4°C overnight or room temperature for 1 hour.

4. Coated plates and flasks can be sealed with Parafilm and stored at 4°C on a level surface for up to 2 weeks.

5. Prior to plating hES cells, aspirate the Matrigel/KO DMEM and rinse with prewarmed KO DMEM.

6. Add hES cells immediately (do not allow Matrigel matrix to dry) in 0.3 ml per cm^2 conditioned hES medium. Plate cells at a density of 2.5×10^4 cells per cm^2.

7. Incubate at 37°C in a humidified incubator with 5% CO_2. Change the medium for fresh conditioned hES medium daily.

8. Human ES cells will require passaging at or just before confluence, beyond which point the medium will be excessively acidified and excess cells will be shed into the medium, giving it a bright yellow, cloudy quality.

TRYPSIN-BASED PASSAGING OF hES CELLS

1. Aspirate the medium from the hES culture and rinse with prewarmed KO DMEM.

2. Incubate the cells in TEG (40 µl per cm^2) at 37°C until all cells have rounded, then knock the flask to release the cells into suspension.

3. Inactivate the trypsin by adding two volumes of prewarmed KO DMEM with 20% FCS.

4. Pellet the hES cells by centrifugation of the cell suspension at 200 **g** for 5 minutes.

5. Resuspend the cells in conditioned hES medium (with bFGF) and plate on a fresh Matrigel-coated surface, rinsed as described above.

6. We recommend reducing the cell density to a half or third of the density of the original confluent monolayer, or plating at 2.5×10^4 cells per cm^2. Use 0.3 ml conditioned medium per cm^2.

NOTES

Protocol 9.2: Preparation of DNA for transfection

1. Isolate supercoiled vector using a quality, column-based maxi-prep kit (Qiagen or Promega). Ensure the DNA is of highest quality by spectrophotometry and gel electrophoresis. Resuspend the DNA at $1\,mg\,ml^{-1}$ in nuclease-free TE buffer or distilled water.

2. For each lipofection prepare $5\,\mu g$ and for each electroporation $50\,\mu g$.

3. For transient transfections (see Protocol 9.3) use supercoiled DNA (go to step 9.2.6); for stable transfections (see Protocols 9.5–9.7), first linearize the DNA (go to step 9.2.4).

4. Select a unique restriction enzyme site within the vector backbone, preferably over 500 bp from any sequences or cassettes required intact.

5. Linearize the vector by incubation in a micro-centrifuge tube at $500\,ng\,\mu l^{-1}$ with $1\,U\,\mu l^{-1}$ restriction enzyme at the manufacturer's recommended conditions for 1 hour. Take a 200 ng DNA sample and check for full linearization by gel electrophoresis.

6. Precipitate DNA by addition of 0.1 volumes of 3 M sodium acetate (pH 4.8) and 2.5 volumes of absolute ethanol. Mix and pellet the DNA by centrifugation at 13 000 rpm in a microcentrifuge for 5 minutes.

7. Wash the DNA pellet by adding 1 ml 70% ethanol. Seal the tube and sterilize the inside by careful inverting, ensuring full contact of internal surfaces with the ethanol.

8. Store the tube unopened at –20°C until required for transfection.

NOTES

Protocol 9.3: Assessment of transfection protocols by transient transfection

This assay relies on transient expression of enhanced green fluorescent protein (EGFP) upon uptake of a mammalian EGFP expression vector into the nucleus of the cell. Transient transfection results in expression in cells that will not stably integrate the vector, thus this assay will give a qualitative comparison of transfection efficiency between different protocols and parameters but will not predict rates of stable transfection. See Protocols 9.5–9.7 for the transfection protocols and select the protocols and parameters to be tested. Conduct comparisons in parallel in a controlled fashion.

1. Prepare sufficient supercoiled pEGFP-C1 (BD Clontech) DNA according to Protocol 9.2 (steps 1–3 and 6–8).

2. Prepare sufficient hES cells by expanding cells on Matrigel matrix in conditioned medium using TEG disaggregation for passaging (see Protocol 9.1).

3. Transfect hES cells according to protocols 9.5–9.7 with the uncut, supercoiled pEGFP-C1 DNA and incubate overnight at 37°C in a humidified incubator with 5% CO_2.

4. After 24 hours, assess the number of live and green fluorescent cells. As numbers of surviving cells can be very low for some electroporation protocols, fluorescence microscopy gives a better indication than FACS analysis.

5. Remove medium and replace with PBS (supplemented with 0.9 mM calcium chloride and 0.5 mM magnesium chloride to prevent rounding of cells).

6. View with a fluorescence microscope (EGFP has an excitation maximum of 488 nm and an emission maximum of 507 nm) and assess the conditions that give the best compromise between low levels of cell death and high levels of cell transfection.

NOTES

Protocol 9.4: Optimization of antibiotic concentrations

1. Coat a 6-well plate with Matrigel.

2. Plate 2×10^5 hES cells in each well and culture in conditioned medium for 48 hours.

3. Apply varying concentrations of the test antibiotic in conditioned medium in five of the wells; continue to culture the sixth in conditioned medium as a control.

4. Change the medium daily with fresh conditioned medium and antibiotic.

5. After 10 days, select the lowest concentration of antibiotic that causes complete cell death.

A range of 50–200 $\mu g\,ml^{-1}$ G418 was found to be effective for H9 and H1 hES cells, although cell death is very slow at the lower concentrations.

NOTES

Protocol 9.5: Lipofection

This protocol is based on the manufacturer's instructions for Lipofectamine 2000 (Invitrogen).

1. Coat two wells of a 6-well plate with Matrigel matrix.

2. Plate 2×10^5 hES cells in each well and culture in conditioned medium for 48 hours.

3. Bring the microcentrifuge tube with a sterile pellet of 5 μg of DNA (see Protocol 9.2) into the tissue culture hood, remove the ethanol supernatant and air dry the pellet under sterile conditions.

4. Resuspend the DNA pellet in 10 μl PBS, add 250 μl Optimem 1 (Invitrogen) and mix.

5. In a separate tube mix 12 μl Lipofectamine 2000 and 250 μl Optimem 1.

6. Combine the two solutions, mix and incubate at room temperature for 25 minutes.

7. Remove the medium from the two wells of cells. Add 3 ml conditioned medium onto one well which will act as an untransfected control. To the second well add 2.5 ml conditioned medium.

8. Add the DNA/Lipofectamine/Optimem mix to the second well and mix gently with the medium in the well.

9. Incubate at 37°C in a humidified incubator with 5% CO_2 for no more than 12–14 hours overnight.

10. Change medium for conditioned medium on both wells. Note: the lipofection reagent will cause a high degree of cell death. Incubate overnight.

11. Change the medium again the following day and incubate overnight.

12. The following day, apply conditioned medium with an appropriate concentration of antibiotic (see Protocol 9.4) to both wells for destruction of untransfected cells.

13. Change the medium daily for fresh conditioned medium with antibiotic.

14. After 10 days to 2 weeks all cells in the control well should have died and any transfected clones will appear as discrete colonies, visible by eye.

15. See Protocol 9.8 for recommendations for expansion of hES colonies.

NOTES

Protocol 9.6: Electroporation using gene pulser

This protocol is based on standard mouse ES cell methods (e.g. see 18) with adjusted parameters to suit H9 hES cells. See Zwaka and Thomson (22) for alternative conditions. For other hES cell lines, test combinations of electroporation parameters (voltage and capacitance) by transient transfection (see Protocol 9.3) to select the best conditions. Both Gene Pulser and Gene Pulser II (BioRad) electroporators are suitable for this protocol. Read the manufacturer's instructions carefully for directions on the safe use of this equipment.

1. Culture hES cells on Matrigel matrix in conditioned medium and passage using TEG (see (22) for alternative).

2. Expand cells to 70% confluence in a 75 cm^2 flask.

3. Place fresh conditioned medium on the hES cells 2 hours prior to transfection.

4. Coat three 15 cm tissue culture dishes with Matrigel.

5. Transfer sterile pellet of 50 μg DNA (see Protocol 9.2) to a tissue culture hood, remove ethanol supernatant and air dry in sterile conditions.

6. Resuspend the DNA pellet in 50 μl HBS.

7. Aspirate the medium from the cells, rinse with prewarmed KO DMEM and aspirate.

8. Incubate cells with 3 ml TEG at 37°C until all cells have rounded, then knock the flask to release the cells into suspension (see (22) for alternative).

9. Inactivate the trypsin by adding 6 ml of prewarmed KO DMEM supplemented with 20% FCS.

10. Remove a small sample and evaluate cell density using a hemocytometer.

11. Pellet the hES cells by centrifugation of the cell suspension at 200 **g** for 5 minutes.

12. Resuspend the cells at 1.33 × 10^6 cells per ml in HBS (see (22) for alternative) at room temperature.

13. Take 375 μl cells and plate in one of the Matrigelled dishes in 25 ml conditioned medium as an untransfected control. Place in the incubator.

14. Place the 50 μl DNA in a 0.4 cm gap electroporation cuvette (BioRad), add 750 μl of cell suspension and mix carefully avoiding bubbles.

15. Electroporate in Gene Pulser II at room temperature using 200 V, 950 μFD (see (22) for alternative) and leave cells to stand at room temperature for 10 minutes.

16. Transfer the cells to 10 ml prewarmed conditioned medium and plate 5 ml to each of the two remaining Matrigelled dishes with 20 ml conditioned medium.

17. Incubate all three plates overnight at 37°C in a humidified incubator with 5% CO_2.

18. Change the medium the next day and incubate overnight.

19. The following day, apply conditioned medium with an appropriate concentration of antibiotic (see Protocol 9.4) for selection of stably transfected clones.

20. Change the medium daily for fresh conditioned medium with antibiotic.

21. After 10 days to 2 weeks all cells in the control plate should have died and any transfected clones will appear as discrete colonies, visible by eye.

22. See Protocol 9.8 for recommendations for expansion of hES colonies.

NOTES

Protocol 9.7: Electroporation using multiporator

This protocol is based on recommendations in the Multiporator Basic Applications Manual (Eppendorf) and has been optimized to suit H9 hES cells. Instructions are given below for optimization for other hES cell lines. Read the manufacturer's instructions carefully for directions on the safe use of this equipment.

OPTIMIZATION

1. Eppendorf supply hypo-osmolar and iso-osmolar buffer. Prepare buffer mixes from 100% hypo-osmolar and 0% iso-osmolar to the reverse. Incubate hES cells in these buffers for 30 minutes at room temperature. Also incubate cells in conditioned medium as a control.

2. Mix 10 μl of cells with 10 μl trypan blue and observe promptly. Discount any buffer conditions that cause greater than 10% of the cells to lyse and take up the stain.

3. Centrifuge the hES cells at 200 g for 5 minutes, resuspend in conditioned medium and culture on Matrigel for 48 hours. Discount any buffer conditions that cause an unusual amount of cell death compared with the control. Select the most hypotonic buffer conditions that were not discounted above.

4. Swell the hES cells in the selected hypotonic buffer conditions for 20 minutes at room temperature. Measure the expanded hES cell diameter by placing a drop of cells on a hemocytometer and comparing with gradations of known size.

5. Estimate the critical field strength (E_c) using the formula:

$$E_c = \frac{V_c}{0.75 \times \text{cell diameter (cm)}} = \text{V.cm}^{-1}$$

where V_c = 1 V at 20°C, 2 V at 4°C.

note: 1 μm = 1×10^{-4} cm.

6. Use a minimal voltage (V_{min}) of $E_c \times 0.4$ for a 0.4 cm gap cuvette. As voltages, try: $1 \times V_{min}$, $2 \times V_{min}$, $3 \times V_{min}$, $4 \times V_{min}$ and $5 \times V_{min}$.

7. Try field lengths of 40, 70 and 100 μS.

8. Test all combinations of voltages and field lengths identified above (15 in total) using the transient transfection assay of electroporation parameters (Protocol 9.3) and select the most effective.

9. If none of these parameters are effective refer to the Multiporator Basic Applications Manual (Eppendorf) for additional suggestions, including higher voltages, multiple pulses and electroporation at 4°C.

ELECTROPORATION

1. Culture hES cells on Matrigel matrix in conditioned medium and passage using TEG.

2. Expand cells to 70% confluence in a 75 cm² flask.

3. Place fresh conditioned medium on the hES cells 2 hours prior to transfection.

4. Coat three 15 cm tissue culture dishes with Matrigel.

5. Transfer sterile pellet of 50 μg DNA (see Protocol 9.2) to a tissue culture hood, remove ethanol supernatant and air dry in sterile conditions.

6. Resuspend the DNA pellet in 50 μl sterile double distilled water.

7. Aspirate the medium from the cells, rinse with KO DMEM and aspirate.

8. Incubate cells with 3 ml TEG at 37°C until all cells have rounded, then knock the flask to release the cells into suspension.

9. Inactivate the trypsin by adding 6 ml of prewarmed KO DMEM with 20% FCS.

10. Remove a small sample and evaluate cell density using a hemocytometer.

11. Pellet the hES cells by centrifugation of the cell suspension at 200 *g* for 5 minutes.

12. Resuspend the cells at 1.33×10^6 cells per ml in the hypotonic buffer conditions identified above and incubate at room temperature for 20 minutes.

13. Take 375 μl cells and plate in one of the Matrigelled dishes in 25 ml conditioned medium as an untransfected control. Place in the incubator.

14. Place the 50 μl DNA in a 0.4 cm gap electroporation cuvette (BioRad), add 750 μl of cell suspension and mix carefully avoiding bubbles.

15. Electroporate in Multiporator at room temperature at the selected voltage (V) and field length (μS) and leave cells to stand at room temperature for 10 minutes.

16. Transfer the cells to 10 ml prewarmed conditioned medium and plate 5 ml to each of the two remaining Matrigelled dishes with 20 ml conditioned medium.

17. Incubate all three plates overnight at 37°C in a humidified incubator with 5% CO_2.

18. Change the medium the next day and incubate overnight.

19. The following day, apply conditioned medium with an appropriate concentration of antibiotic (see Protocol 9.4) for selection of stably transfected clones.

20. Change the medium daily for fresh conditioned medium with antibiotic.

21. After 10 days to 2 weeks all cells in the control plate should have died and any transfected clones will appear as discrete colonies, visible by eye.

22. See Protocol 9.8 for recommendations for expansion of hES colonies.

NOTES

Protocol 9.8: Expansion of hES colonies

Once human ES colonies appear and untransfected cells have died the hES clones can be expanded without further antibiotic selection, either on feeder layers or, as described below, on Matrigel matrix in conditioned medium. The following protocol has been successful for H1 and H9 hES. If using other human ES cell lines, first plate cells at clonal density to generate some colonies and test this method. If cells do not survive, it may be necessary to isolate the colonies by 'ring' or 'tower' cloning using the disaggregation reagent most suitable for that cell line.

1. Coat a 48-well plate with Matrigel, one well for each colony to be isolated. Place 450 µl of conditioned hES medium in each well.

2. Aspirate the medium from a plate of colonies, wash with PBS (Ca^{2+} and Mg^{2+} free) and aspirate.

3. Place another 20 ml fresh PBS on the plate and incubate at room temperature for 5–10 minutes until the hES cells can be seen to be rounding.

4. Mechanically isolate each colony with a micropipette and tip by scraping it and drawing it up in a 50 µl volume. This should be possible by eye but can be performed on a dissection microscope in a tissue culture hood.

5. Place each colony into a well with conditioned medium and leave the plate undisturbed for 24 hours at 37°C in a humidified incubator with 5% CO_2.

6. The cells will attach and grow as clumps of cells. Change the medium daily until the clumps become very large and dense.

7. Rinse wells with PBS and treat each with 100 µl TEG. Incubate until cells detach.

8. Pipette cells to suspend and transfer into a Matrigelled 24 well with 0.5 ml conditioned medium.

9. Incubate and change medium daily until confluent.

10. At this stage cells can be passaged into duplicate 12-well plates by treating PBS-washed cells with 300 µl TEG, transferring 100 µl and 200 µl to duplicate Matrigelled 12-well plates containing 1 ml of conditioned medium.

11. Incubate and change medium daily until confluent.

12. A confluent 12 well of hES cells is sufficient for cryopreservation, or to lyse for preparation of genomic DNA for Southern blotting or PCR analysis.

13. For further expansion of the cells, continue to passage each time onto a tissue culture surface with two or three times the original surface area, until the desired culture size is reached.

NOTES

Index